Wei Tian

Novel Aggregated Solutions for Robust Visual Tracking in Traffic Scenarios

Schriftenreihe
Institut für Mess- und Regelungstechnik,
Karlsruher Institut für Technologie (KIT)

Band 044

Eine Übersicht aller bisher in dieser Schriftenreihe erschienenen Bände finden Sie am Ende des Buchs.

Novel Aggregated Solutions for Robust Visual Tracking in Traffic Scenarios

by
Wei Tian

KIT Scientific Publishing

Dissertation, Karlsruher Institut für Technologie
KIT-Fakultät für Maschinenbau

Tag der mündlichen Prüfung: 31. Januar 2019
Hauptreferent: Prof. Dr.-Ing. Christoph Stiller
Korreferent: Prof. Dr.-Ing. Fernando Puente León

Impressum

KIT Scientific Publishing

Karlsruher Institut für Technologie (KIT)
KIT Scientific Publishing
Straße am Forum 2
D-76131 Karlsruhe

KIT Scientific Publishing is a registered trademark
of Karlsruhe Institute of Technology.
Reprint using the book cover is not allowed.

www.ksp.kit.edu

Print on Demand 2019 – Gedruckt auf FSC-zertifiziertem Papier

ISSN 1613-4214
ISBN 978-3-7315-0915-8
DOI 10.5445/KSP/1000091919

Novel Aggregated Solutions for Robust Visual Tracking in Traffic Scenarios

Zur Erlangung des akademischen Grades eines

Doktors der Ingenieurwissenschaften (Dr.-Ing.)

von der KIT-Fakultät für Maschinenbau des

Karlsruher Instituts für Technologie (KIT)

angenommene

Dissertation

von

M.Sc. Wei Tian

Tag der mündlichen Prüfung:	31.01.2019
Hauptreferent:	Prof. Dr.-Ing. Christoph Stiller
Korreferent:	Prof. Dr.-Ing. Fernando Puente León

Prologue

This present thesis was written during my time as a scientific employee at the Institute of Measurement and Control Systems at the Karlsruhe Institute of Technology (KIT). It was completed under the supervision of Prof. Dr.-Ing. Christoph Stiller. I would like to express great thanks to him for offering me the opportunity, as a Ph.D. student with enthusiasm, to explore scientific areas such as computer vision and automated driving at a high quality level of current researches as well as providing a reliable research environment and a strong research team. I would also like to thank my co-supervisor Prof. Dr.-Ing. Fernando Puente León for the supervision on my thesis and interest in my work.

Additionally, I would like to express my sincere thanks to my team leader, Dr. Martin Lauer. In our regular meetings, he provided valuable thoughts and advices, which ensured a persistent guidance on the path of my research and have enlightened the ideas of my papers. I would also like to thank the Ph.D. examination committee of Department of Mechanical Engineering for their permission of my pursuit of a Ph.D. degree. I am also grateful to Karlsruhe School of Optics and Photonics (KSOP) for providing the financial support for my first year.

Furthermore, I would also like to give my thanks to my dear colleagues (also including the guest researchers). They provided a great support to me, giving a lot of helpful suggestions and interesting comments about my research. It is the great honor in my life to work together with them and share almost five wonderful years.

This thesis is based on several research works published during my period as a Ph.D. student. Most of these works are based on the collaboration within research groups of my institute. Without their support and help, these works and this thesis would not have been possible. In gratitude of their contributions as well as to make this thesis more reader friendly, the plural form of pronoun, i.e., the word “we”, is used as a multi-person perspective for the narrative throughout this thesis.

Karlsruhe, in January, 2019 *Wei Tian*

Abstract

Technologies related to advanced driver assistance systems and automated driving vehicles have enjoyed a rapid development in recent years. In these highly automated systems, visual tracking approaches are widely employed to analyze the behavior of traffic participants, e.g., pedestrians and vehicles, in the surrounding environment to provide reliable information for functions such as motion control, maneuver determination and collision avoidance. Despite tremendous progress achieved, existing tracking approaches still have difficulties dealing with challenging scenarios like severe occlusion, deteriorated vision and long range multi-object reidentification.

To address above mentioned problems, in this thesis, novel tracking solutions are presented, which aggregate information in levels from visual features to object parts/groups. All these solutions are only based on image sequence captured by a monocular camera and do not require additional sensors. To track a severely occluded object, a part filter-based tracker is employed, in which the occurrence of occlusion is recognized through the variation of the appearance model and the classifier response. The part filter is only learned on the visible object area identified in pixel-level precision by a masking process and is demonstrated with high robustness in experiments. For handling deteriorated vision, a new tracker is presented, which decomposes features into several expert filters and searches the most discriminative one based on their estimated reliabilities. Additionally, it performs an optimization in the temporal domain to filter out corrupted samples. Both procedures are integrated in a single learning scheme and the trained tracker yields favorable performance in cases with low illumination or adverse weathers. Following the trending technique of tracking-by-detection and leveraging the advances in object detection, multi-object tracking can be cast as a reidentification/association problem. To efficiently process large amount of objects, a three-stage association scheme is presented in this thesis, which is mainly based on the strategy of joining both spatial and temporal constraints. The first one encodes the relative motion between targets while the second one focuses on long ranged objects. Such frameworks can cope with both camera motion and full occlusion. Integrated with previously introduced trackers, it exhibits an improved state-of-the-art performance in more challenging scenarios. Since all the presented approaches are carefully designed, they run at a real-time speed.

Kurzfassung

Technologien im Feld von Fahrerassistenzsystemen und autonomen Fahrzeugen haben in den letzten Jahren eine rasante Entwicklung erfahren. In diesen hochautomatisierten Systemen werden visuelle Tracking-Ansätze häufig verwendet, um das Verhalten von Verkehrsteilnehmern, z.B. Fußgängern und Fahrzeugen, in der Umgebung zu analysieren sowie um zuverlässige Informationen für Funktionen wie Bewegungssteuerung, Manöverplannung und Kollisionsvermeidung bereitzustellen. Trotz enormer Fortschritte haben bestehende Verfolgungsansätze immer noch Schwierigkeiten mit anspruchsvollen Szenarien wie starker Verdeckung, verschlechterter Sicht und der Wieder-Identifizierung mehrerer Objekte.

Um die oben genannten Probleme zu lösen werden in dieser Arbeit neuartige Tracking-Ansätze vorgestellt, die Informationen in Ebenen nutzen von visuellen Merkmalen bis hin zu Objektteilen/-gruppen. All Lösungen basieren nur auf einer Bildsequenz, die mit einer monokularen Kamera aufgenommen wurde, und erfordern keine zusätzlichen Sensoren. Um ein stark verdecktes Objekt zu verfolgen, wird ein Part-Filter-basierter Tracker verwendet, bei dem das Auftreten der Verdeckung durch die Variation des Aussehen-Modells und der Klassifiziererantwort erkannt wird. Der Part-Filter wird nur auf dem sichtbaren Objektbereich gelernt, der durch einen Maskierungsprozess in pixelweiser Genauigkeit identifiziert wird. Seine hohe Robustheit wird in Experimenten demonstriert. Um die verschlechterte Sicht zu behandeln, wird ein neuer Tracker vorgestellt, der die Merkmale in mehrere Expertenfilter zerlegt und die diskriminierendsten Filter anhand ihrer geschätzten Zuverlässigkeit durchsucht. Zusätzlich führt es eine Optimierung in der zeitlichen Domäne durch, um beschädigte Samples herauszufiltern. Beide Verfahren sind in einem einzigen Lernschema integriert, und der trainierte Tracker liefert eine günstige Leistung in Fällen mit geringer Beleuchtung oder widrigen Wetterbedingungen. Nach dem Tracking-by-Detection-Verfahren und dank des Fortschritts bei Objektdetektionstechniken kann das Multiobjekt-Tracking als ein Assoziationsproblem gedeutet werden. Um eine große Anzahl von Objekten effizient zu bearbeiten, wird in dieser Arbeit ein dreistufiges Assoziationsschema vorgestellt, das hauptsächlich auf der Strategie basiert, sowohl räumliche als auch zeitliche Bedingungen zu kombinieren. Die erste kodiert die relative Bewegung zwischen Objekten, während die zweite sich auf Objekte in großer Entfernung konzentriert. Ein solcher Rahmen kann die Kamerabewegung und langzeitig voll verdeckte Objekte gut behandeln. In Kombination mit den bereits vorgestellten Trackern bietet es eine verbesserte Leistung auf dem neuesten Stand der Technik in anspruchsvolleren Szenarien. Da alle vorgestellten Ansätze elegant gestaltet sind, erlauben sie auch eine Echtzeitgeschwindigkeit.

Contents

Notation and Symbols

Acronym

2-D/3-D	**2/3-D**imensional
ACC	**A**daptive **C**ruise **C**ontrol
ACS	**A**lternate **C**onvex **S**earch
AD	**A**utomated **D**riving
ADAS	**A**dvanced **D**river **A**ssistance **S**ystem
AED	**A**verage **E**uclidean **D**istance
AL-ICM	**A**daptive **L**abel **I**terative **C**onditional **M**odes
BIP	**B**inary **I**nteger **P**rogramming
CF	**C**orrelation **F**ilter
CNN	**C**onvolutional **N**eural **N**etwork
DFT	**D**iscrete **F**ourier **T**ransform
HDR	**H**igh **D**ynamic **R**ange
IR	**I**nfra**r**ed
KCF	**K**ernelized **C**orrelation **F**ilter
LKT	**L**ucas-**K**anade **T**racker
MCMC	**M**arkov **C**hain **M**onte **C**arlo
MOTA	**M**ultiple **O**bject **T**racking **A**ccuracy
MOTP	**M**ultiple **O**bject **T**racking **P**recision
MT	Ratio of **M**ostly **T**racked Targets
ML	Ratio of **M**ostly **L**ost Targets
IDS	Number of **Id**entity **S**witches
FR	Number of Track **Fr**agments
NOD	**N**ormalized **O**bject **D**ifference
PSR	**P**eak-to-**S**idelobe **R**atio
SLAM	**S**imultaneous **L**ocalization **a**nd **M**apping
SVM	**S**upport **V**ector **M**achine
VRU	**V**ulnerable **R**oad **U**ser

General Notation

Scalars	Regular lower case	a, b, c
Vectors	Bold lower case	$\mathbf{a}, \mathbf{b}, \mathbf{c}$
Matrices	Bold upper case	$\mathbf{A}, \mathbf{B}, \mathbf{C}$
Sets	Bold italic upper case	$\boldsymbol{A}, \boldsymbol{B}, \boldsymbol{C}$

Operation

$\hat{\mathbf{x}}$	Fourier Transform of vector **x**
$\mathbf{x}^H$	Hermitian transpose of vector **x**
$\mathbf{a} * \mathbf{b}$	Convolution between vector **a** and **b**
$\mathbf{a} \odot \mathbf{b}$	Hadamard product between vector **a** and **b**
$f(\mathbf{z})$	Response score on sample **z**
$\mathbf{f}(\mathbf{z})$	Response map on sample **z**

1 Introduction

According to statistics [123], in last decade, the vehicle ownership has been significantly increased and the number of vehicles in operation world widely has reached 1.32 billion in 2016. However, the rapid increase in vehicle number not only improves the mobility for the vast majority of people but also brings challenging issues, such as traffic congestions and accidents, which result in enormous social burdens. Only in the year of 2016, the average congestion time is about 30 hours on German roads [33], which causes a total cost of over 69 billion euros. In the meanwhile, road accidents become one of the major threats to the safety of human beings [163]. Over 56.6 million vehicles crashed on German roads in 2016 [14]. The associated cost is estimated around 34 billion euros and over 90% of traffic accidents are caused by human factors.

To alleviate congestions on road and to reduce the amount and severity of vehicle accidents, both automobile industries and research institutions have made great endeavors to upgrade the technology. Representatively, the advanced driver assistance systems (ADAS) are applied in tasks such as congestion assistance, adaptive cruise control (ACC), obstacle avoidance, collision warning, etc., to help the driver make maneuver decisions and recognize objects in the surroundings. Such developments led to a pronounced improvement on the quality of traffic flow [80] and a steady decline in car accidents [14].

As a prominence of intelligent transportation systems, the automated driving (AD) is regarded as the future automobile technology, which aims to significantly improve the traffic safety as well as efficiency and comfortability. The automated driving system is able to partially or fully take over the control of the vehicle especially in critical scenarios. Thanks to advances from sensing and computing technologies, such system operates much more robust and is insusceptible to human-related factors such as distraction, fatigue, drowsiness and other dangerous driving behaviors. In the meanwhile, it also saves human drivers a lot of time and energy by freeing them from heavy driving burdens.

One of the most essential technologies for automated driving is the environment perception, which persistently monitors various objects and events in complex traffic scenarios and provides vital information for sub-processing modules such as localization, path planning, motion control and maneuver decision. To ensure both a high reliability and a high accuracy, the perception system is usually implemented by heterogeneous sensors which are able to generate high-resolution, high-frequency data and with lots of computation power and a wide communication bandwidth. Currently, a vast amount of sensor setups are available off-the-shelf and applied in the development of automated driving systems. For instance, one of the perception systems from Google is basically a powerful multi-line lidar in combination with additional camera and radar sensors [62]. However, the Tesla Model S mainly utilizes a sensor setup of eight monocular cameras aided with radar and ultrasonic sensors [56]. A similar setup is also adopted by Daimler yet enhanced with stereo and infrared vision [152]. Although the sensor selection differs for each developer, the camera-based visual system is still a common choice for the main perception solution. Such systems can provide remarkable discriminative power and are utilized to recognize objects like traffic signs and lane markings, and to analyze the trajectory or even behavior of vehicles and pedestrians. In the latter case, the related technology is the tracking approach, which is usually implemented in vision systems and also the focus of this thesis.

1.1 Vision System and Tracking Approach

The vision system is an essential perception technology with interdisciplinary applications. In biology, especially for human beings, it refers to the eyes, which perceive about 80% of all environmental information [77], far more than other organs of perception. For artificial systems, we mainly talk about the camera-based perception devices. They are able to capture the shape, color and texture information, which are difficult to obtain by radar or ultrasonic sensors [20] but are requisite features for interpretation of objects or scenes. Thanks to the advances from the domain of optics and photonics, both camera size and weight have been significantly reduced in recent years while their performance persistently increased. In the meanwhile, both the manufacturing cost and market price of cameras have been significantly reduced due to mass production. All these points led to the popularity of camera sensors and made it possible to integrate them into small and flexible mobile systems like smart phones and specific devices in automated driving vehicles.

Generally, a vision system can be decomposed into two main parts [77]: the imaging component like optics to project the scene and the processing component to parse the image information. While the first part is mainly related to hardware, the second part emphasizes more on software levels. Normally, domain knowledge from computer vision is leveraged to extract and analyze image information, making camera-based perception systems suitable for various applications. As illustrated in Fig. 1.1 (a), for blind spot surveillance, side cameras are installed on the vehicle to identify pedestrians or cyclists and analyze their behaviors to prevent possible collisions [148, 149]. The behavior analysis can be extended in 3D vision by stereo setups or additional range sensors [61]. In other applications, camera systems are installed in the cabin to analyze the eye movements of the driver (Fig. 1.1 (c)) and identify distraction and drowsiness [76]. Aside from that, the motion of objects can also be estimated by analyzing the position change of points over images [67], which is illustrated in Fig. 1.1 (b). Along the resurgence of deep learning in recent years, neural networks are introduced into computer vision and image processing domain. Depending on huge amounts of training data and high performance processors, a lot of problems which are difficult for traditional approaches become tractable by neural networks, especially in object identification, action recognition (Fig. 1.1 (d)), scene understanding, etc. For specific areas, neural networks even surpass the performance of human beings [121]. Thus, they become more popular in the development of ADV technologies. In above discussion, although applications of vision system differ under circumstances, most of them are related to the analysis of object behaviors in observed scenes. The key information is the object trajectory which is mostly interpreted in image coordinates and derived by tracking approaches. Thus, vision-based tracking becomes an indispensable part of many perception systems.

Vision-based Tracking Approach

Object tracking is commonly considered as estimating the trajectory of an object by analyzing its movements in a predefined time interval. Before the broad application of camera-based vision systems, targets were tracked mainly by radar sensors in precedent works [2, 157]. Due to low resolution of the sensors, objects are normally considered as rigid points, which are identified by correlating signals with their echos or by deriving the posteriori density of the state using Bayesian approaches. However, with such a poor feature interpretation, they are unable to deal with scenarios such as to distinguish objects which are, with similar motion patterns, in a very close distance.

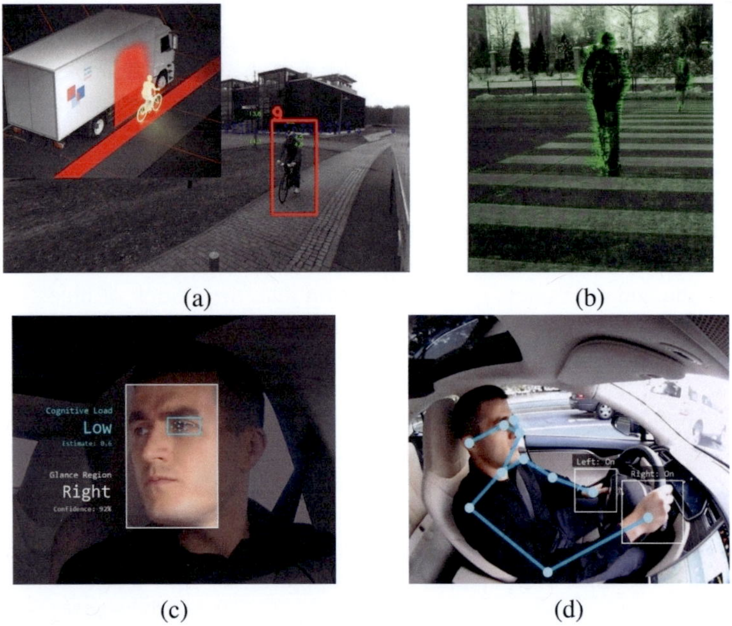

Figure 1.1: Applications of vision system. Image (a) to (d) show collision surveillance [61], motion estimation [67], eyeball tracking and action recognition [52], respectively.

By applying vision systems such as cameras, the problem of interpreting the appearance of an object is significantly alleviated by the rich image information. Thus, the tracking task is cast into finding the object with the most similar appearance of the target. Since objects are observed in the image plane, their estimated trajectories are also represented by image coordinates. This methodology is called visual tracking.

The visual information can also be adopted for point tracking and one of its successful applications is the optical flow [55], which interprets motion of points on an object or in a scene over consecutive images. Besides the analytical method which solves a set of differential equations [102], the optical flow also employs block matching to measure the shift of small image regions. The query point is usually centered in the small block region and visual information such as color and texture are encoded by descriptors. Thanks to enormous progress on both descriptor structures like SIFT [101], SURF [15], FAST [130], BRIEF [22] and matching operations which are conducted pyramid-wise [13] or embedded in deep networks [158], optical flow is widely employed by ADV technologies, e.g., in sensor calibration, visual odometry and simultaneous localization and mapping (SLAM).

Apart from points, complex objects such as pedestrians, vehicles, etc., are also considered as tracking targets in various applications. Numerous related approaches have been proposed in recent years and the main progress has been achieved in areas of appearance representation and object reidentification. To represent the appearance of an object, features are constructed in different manners, e.g., modeling the shape with blobs [164], seeking distinctive contours [78], extracting color or gradient histograms [31], ensembles of templates of feature points [153], etc. Besides hand-crafted features, some research works [105, 114] also adopted convolutional neural networks (CNNs) to learn the object appearance. Since most of these CNNs are pre-trained on huge image data sets, the extracted deep features are learned to capture the most distinctive information of objects.

The simplest approach for target reidentification is matching, which can be interpreted as a correlation operation or as a distance metric such as Euclidean, Mahalanobis, etc. Inspired by the image classification technique, the trend of tracking-by-detection surfaces among the tracking community. In such trend, machine learning techniques like boosting, random forests and even CNNs are adopted to train classifiers to distinguish the target from background objects. The correlation filter is one of the most successful approaches. It is derived from the structural support vector machine (SVM) and due to its high performance, in terms of both precision and computation efficiency, it became a hot research topic [166, 167].

Unlike single object tracking, a simple appearance matching approach is usually insufficient to track multiple objects. The most challenging issue for multi-object tracking is the data association problem. Due to the fact that objects are allowed to enter and exit the scene observed by the camera and the motion of each object may differ from each other, the number of objects in the image may vary over the time. Ambiguity between detections and tracked targets occurs when objects are close in position or similar in appearance. The problem of multi-object tracking has been studied for almost a half century [2]. Recent successful works are mostly based on a combination of object appearance matching, motion modeling and additional association strategies [126, 128, 142].

Problems and Solutions

Despite tremendous progress achieved in object tracking techniques and exciting performance of newly developed trackers reported by standard benchmarks [90, 166, 167], corner cases for visual tracking such as severe occlusion, deteriorated vision and multi-object reidentification still exist. Coping with these challenges is considered difficult, even for state-of-the-art tracking approaches, especially with monocular camera images only. Thus, these challenges are stumbling rocks on the way to ADV technology, which should ensure a high reliability and safety for the vehicle in nearly all possible traffic situations. In the context of traffic scenes, we address three main challenges for visual tracking in this thesis, which are listed as follows:

- *Severe Occlusion*
  This scenario usually occurs for tracking pedestrians or cyclists, which are considered as vulnerable road users (VRUs). Since their sizes are relatively small, they are easy to be occluded by big objects such as vehicles on the street and only leave a small body part exposed (see Fig. 1.2 (a)-(b)). However, most of state-of-the-art trackers cannot recognize the occurrence of occlusion or have difficulties in identifying the non-occluded object part. This can lead to loss of target in the tracking approach and further to failures of the collision warning system, yielding a dangerous situation for VRUs, since the driver barely has any time to react before an accident happens.

- *Deteriorated Vision*
  Here it distinguishes between two cases: nearly constant and temporally varying vision deterioration. In the first case, it is mainly caused by poor illumination conditions, such as dark scenarios like in the night or

whitened scenarios like in the fog. Driving in these scenarios, visual features in the image are significantly degraded and only some distinctive parts of an object, e.g., the vehicle lamps or reflection stripes, can be observed (Fig. 1.2 (c)). Such cases require the tracker to be able to seek the most discriminative visual feature of an object, which is yet not the case for most standard trackers. The second case is mostly related to the impact of weather conditions like the rain. As illustrated in Fig. 1.2 (e), raindrops falling on the windshield of the vehicle result in an unclear vision of the camera mounted directly behind it. Although the windshield wiper clears the windshield afterwards, the unclear image is still saved in the storage, which deteriorates the training samples of the tracker, further leading to tracking drifts or failures (Fig. 1.2 (f)).

- *Multi-object Reidentification*
  This reidentification problem frequently occurs for tracking multiple objects. In the tracking-by-detection paradigm, association ambiguity appears when the object number varies over the time, which is due to the object-scene-interaction (e.g., objects enter or exit the scene) or mutual occlusion between objects. In the latter case, especially when objects are fully occluded and disappear for a long time, most of standard tracking approaches consider these objects as "dead" due to their long term disappearance and terminate the estimation for their trajectories. Even though these objects may be rediscovered after the occlusion, they will be assigned with different identities due to failed association (Fig. 1.2 (h)). Such a case is called as "ID switch", which is a common, not well-solved problem among the tracking community.

Since the above mentioned corner cases are greatly influenced by environment factors or by crowd behaviors, current tracking approaches which mostly consider individual targets within non-complex situations can only yield inferior performance. To tackle these challenging issues, the key point for a tracking approach is to appropriately manipulate object models, channel features, training samples and even association strategies under specific circumstances. In this thesis, we call such approaches aggregated solutions.

The terminology of "aggregation" stems from statistics and refers to the compiling of information from combined datasets with intent to prepare them for data processing [140]. The concept is broadly applied in data science domain.

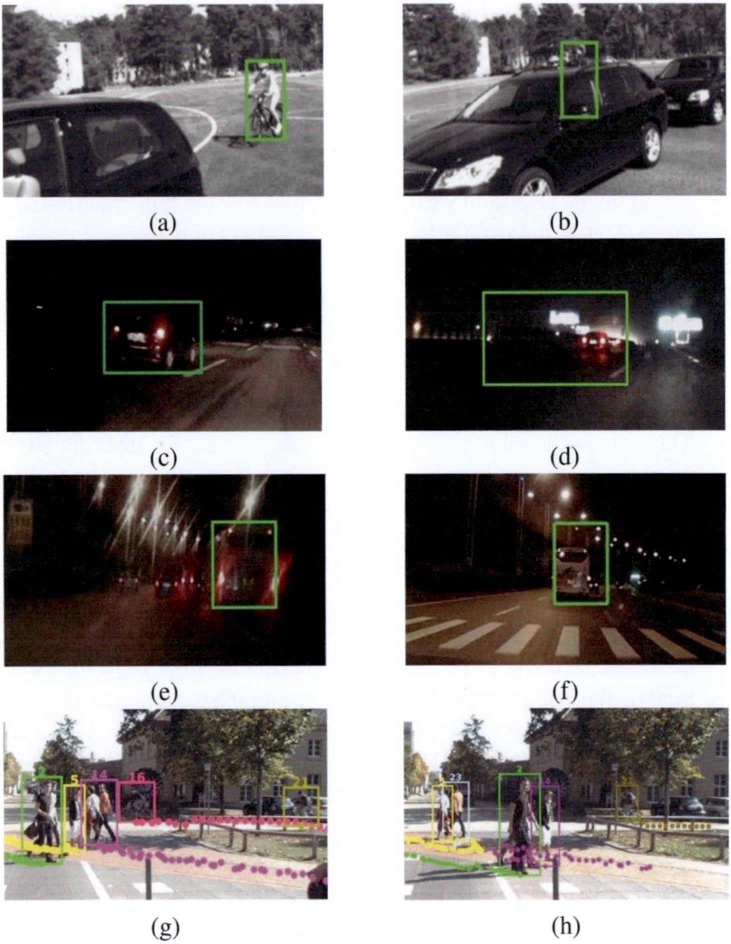

Figure 1.2: Challenges for visual tracking. In each row it respectively shows a tracking scenario with severe occlusion, constantly deteriorated vision, temporally deteriorated vision and multi-object reidentification. In the first three scenarios, the utilized tracker is a correlation filter and the tracking failure or drift is displayed in the second column. The last scenario is for multi-object tracking. The association is conducted by the Hungarian algorithm [91], which is adopted by a vast of research works. In the next image, it shows the failed association, i.e., ID switch, between object 5 and object 14.

It is used to optimize the collecting and transfer of data from distributed systems [63, 137] as well as in machine learning to manipulate training datasets (e.g., bootstrapping) and to learn joined classifiers (e.g., ensemble learning). Here we introduce this concept in the visual tracking domain and extend its definition to cover any manipulation related to compiling information with different resources to improve the tracking performance. Thus, it ranges from low levels, e.g., features and samples, to high levels such as object parts and crowds. In this sense, proposed solutions in this thesis can also be considered as aggregated approaches which work on different levels to solve each of the above mentioned challenging scenarios.

1.2 Contributions

In this thesis, novel aggregated approaches are proposed to solve challenging issues for visual tracking in traffic scenarios like severe occlusion, deteriorated vision and multi-object reidentification. All proposed approaches can work with monocular camera images and require no further information from additional sensors. Contributions of this thesis are summarized as follows:

- To deal with severe occlusion, a new tracking approach by adopting part filters is proposed. In this approach, the occluded parts of an object are recognized by utilizing the information derived from both image features and filter responses. By a masking procedure, visible object areas can be obtained with pixel-level accuracy and used to build part filters. By experimental results on real traffic scenes, the tracker is proven to be vigorous against occlusion, particularly in cases where long term severe occlusion happens. It is also demonstrated by further experiments on a standard benchmark that this approach improves object tracking performance under other challenging circumstances. This approach has been published in works [146], [151].
- For deteriorated vision, a novel online method is proposed to manipulate features and samples in the learning procedure to adjust the tracker to environmental factors. In this method, channel features are assigned to various experts. Appearance models of an object are constructed based on their evaluated reliabilities to emphasize the most discriminative

visual features, which can handle constant deteriorated vision, such as low illumination conditions during the night. Additionally, a temporal optimization is performed to suppress outliers and maintain the most reliable samples to train the tracker, which thus becomes insusceptible to temporally varying vision contamination such as in rainy weather. An outstanding performance of this approach is exhibited by experiments on real driving datasets with scene tags of night, fog and rain. The related work is published in [144].

- A novel multi-object tracking approach is proposed to address the reidentification issue for multiple targets observed in large spatial and temporal domains. In this approach, detections are firstly amassed into short tracklets according to affinity measurements. In a short time span, motion patterns and spatial relationships within grouped targets are used to link tracklets to existing objects. In a period window of bigger scale, graph theories are adopted to recover objects, which vanished for quite a long time due to failed detections or long-term occlusions. Leveraging both strategies, this approach exhibits an improved state-of-the-art performance, demonstrated in works [104, 145, 147, 175] [1].

1.3 Thesis Outline

In this thesis, we address aggregated solutions in terms of different challenging scenarios of visual tracking. Thus, this thesis consists of topics from different perspectives. For a better narrative, only a general overview about the development of vision system and tracking approach has been placed in the introduction of this thesis. Details about the state of the art of each topic is given in the beginning of each chapter.

Hence, the remainder of this thesis is organized as follows: In **Chapter 2** we give a brief introduction about the correlation filter, which is the basic framework for most of our proposed approaches. We introduce its development, its theoretical derivation and the specific versions utilized in our approaches.

[1] In works [144–147, 151], the author of this thesis contributed to most of the ideas and most of the writing work.

Chapter 3 describes the proposed approach which adopts part filters to solve tracking with severe occlusions. At the beginning it is about occlusion awareness, which is divided into two parts: occlusion occurrence detection and occluded object part identification. Afterwards we move to dynamic filter management and finally to the evaluation results. In **Chapter 4** we respectively describe the approaches for object tracking with constant and temporally varying vision deterioration. We give the detailed mathematical derivation and show how to join two sub-approaches into one tracking framework. Finally, we evaluate the approach on a challenging dataset. In **Chapter 5** we describe our multi-object tracking approach and introduce two association strategies. The first one is based on spatial relationships and motion patterns within grouped targets and is called as spatial constraint. The second one is based on graph-cutting and is utilized in a large-scale time domain. Thus, it is called temporal constraint. We join two constraints in one tracking framework and evaluate it on several standard benchmarks. In **Chapter 6** we give the conclusion of this thesis and end it with research directions in the future work.

Chapter	Topic of Tracking	Aggregation Level	Base Technique
2	Correlation Filter	-	Support Vector Machine
3	Severe Occlusion	Object Parts	Correlation Filter
4	Vision Deterioration	Feature, Samples	Correlation Filter
5	Multi-object Reidentification	Objects	Graph Cutting

Table 1.1: An overview of the topic, the information aggregation level and the base technique of each chapter except the introduction and conclusion part of this thesis.

An overview about the relationship between the main chapters of this thesis is also given in Table 1.1. As illustrated, except Chapter 5, all other chapters are based on the correlation filter technique and concentrated on the aggregation level of features, samples and object parts. Since they are more related to learning the appearance model for one target, we mainly focus on single-object tracking in these chapters. The discussion about their application for multiple targets and issues like the multi-object association are left for Chapter 5, in which we focus on a higher information aggregation level: the behavior between objects.

2 Basic Framework – Correlation Filter

The discriminative tracking approach commonly leverages machine learning techniques to build a classifier based on given image samples to distinguish the target from its background. However, to ensure a high accuracy, traditional classification methods usually require a large amount of training samples, e.g., extracted from different image locations. Such a sampling procedure is extremely time consuming and brings inevitable computational burden while processing these samples. To tackle this bottleneck, the correlation filter emerges, which employs a circular shift to replace the normal image translation and is learned as a support vector machine, which can be efficiently solved in frequency domain. By incorporating additional techniques such as the kernel function, the correlation filter shows a comparable or even superior performance in comparison with other traditional tracking methods.

Due to the above reasons, the correlation filter is chosen as the basic framework for most of proposed approaches, which are described in Chapter 3 and Chapter 4 of this thesis, respectively. In this chapter, we first give a short review about the development of single-object tracking which includes the discriminative tracking approach and the correlation filter. Thereafter, we present the detailed formulation of the correlation filter from both machine learning and signal processing perspectives. After that, we introduce the specific version of the correlation filter utilized in our approaches.

2.1 State of the Art

For comprehensive studies about the topic of visual tracking the readers can be referred to recent surveys such as [88, 138, 167]. Here we provide a brief overview about the works related to single-camera, single-object tracking, which, depending on how the tracker model is constructed, can be coarsely divided into two categories: generative and discriminative approaches.

Generative Models

Generative tracking approaches typically focus on building an appearance model to represent the target and utilizing it to search the most similar image region in new frames. Since the searching procedure is usually guided by probability inferences or by distance metrics, a powerful visual representation of the target becomes the key to these tracking approaches. For instance, Ross et al. [129] represent the target with a subspace model consisting of eigenvectors from the covariance matrix of image samples. By a maximum likelihood estimation, the location of the target is estimated in consecutive frames. Eigenspace representation is also utilized by Black et al. in [17] but enhanced with a multi-scale and coarse-to-fine matching strategy. In the spirit of brightness consistency, they managed to track an object over long video sequences in which the target undergoes camera motion and view changes. Comaniciu et al. in [32] directly model the object appearance with its probability distribution of color values. The object is also spatially masked by an isotropic kernel to avoid large response variation for adjacent locations on image lattice.

Aside from naive image statistics, other visual representation such as sparse coding is also utilized. Wibowo et al. in [162] calculate sparse coefficient vectors from small patches. These sparse features are utilized to measure the likelihood of an observation and in collaboration of particle filter, it is able to estimate the a posteriori probability of the target on evaluated locations. In another way, the sparse representation of an image sample in [108] is obtained through the L1-regularized least square minimization. The probability of the sample is thus calculated with respect to the reconstruction error of the target. By investigating the trivial coefficients from the minimization, occlusion can also be detected. This work is further developed in [109]. By introducing lower error bound and two-stage reconstruction, the computational efficiency is significantly boosted. The low-rank sparse representation is also investigated in [173], which incorporates background samples in the dictionary and a sparse error term to address occlusion and tracking drift problems.

Unlike them, visual features like superpixels are utilized in [119] in combination with a probabilistic model implemented by Earth Movers's Distance to deal with non-rigid objects. In spite of the fact that both the precision and runtime speed of generative models have been greatly improved by above mentioned techniques, their tracking performance strongly depends on specific probabilistic models or distance metrics, and an appropriate selection of them is still difficult. Another issue of generative models is that they rarely take background image information into consideration, which make their approaches prone to background clutter.

Discriminative Models

Due to the recent progress of machine learning in the domain of image classification, speech recognition, medical diagnosis, etc., tracking-by-detection has risen as one of the most successful paradigms. As one of its representatives, the discriminative approach treats tracking as a classification problem, which learns a classifier to predict the location of the target in a new frame. Since background information is involved in negative training samples, the target can be readily distinguished from other objects in the background.

In prior works, heterogeneous classifiers are adopted by discriminative models. Saffari et al. in [132] combine random forests and online bagging in one scheme. Since it allows consecutively growing new trees, their method shows resistance against occlusions. As a simplified version, random ferns are utilized by Rao et al. in [125]. By incremental learning, their tracker model can be readily adapted to the variation of target appearance. Similar classifiers are employed in [154] but enhanced with boosting, which is demonstrated with a promoted precision in recognition of objects undergoing rotation and view changes. The boosting approach is also deployed for ensemble learning of classifiers in [6]. Their weak classifier is based on the distribution of low level features and the new position of target is identified by mean-shift, which works well with color, gray and infrared (IR) images. In [58], Grabner et al. apply online AdaBoost on features such as Haar [155], HOG [35] and LBP [117] and use the confidence map to estimate target positions, achieving boosted performance. This work is further developed in [60] by formulating the tracker update process in a semi-supervised fashion to reduce the risk of tracking failure or drift. Such a problem is also addressed by Babenko et al. in [7], where they introduce multiple instance learning to create sample bags and train the classifier with high potential positive samples to alleviate the model degradation by target location errors.

Aside from ensemble methods, classification techniques such as nearest neighbors are also utilized, e.g., in [165], where the object is evaluated with respect to its distance to a subspace spanned by selected training samples. Since deep learning becomes a trend in computer vision, neural networks are adopted in plenty of research works. For instance, Jin et al. in [75] train a light weight network augmented with a radial basis function classifier and prove its stability against dramatical changes of object appearance. Nam et al. in [114] train two CNNs to switch the tracking between short and long term modes, which benefits greatly in rediscovering lost targets. Although these above-mentioned methods yield outstanding tracking performance, most of their classifiers require a large amount of training samples, which results in huge consumption of memory and computing resources.

As a remedy of this shortcoming, Hare et al. in [66] build a classifier based on the principles of the structured SVM [14]. By augmenting class labels with location information, its computation load is relatively small, compared with other methods. Such an idea inspires the work [18], where Bolme et al. reformulate the learning procedure of SVM in frequency domain, significantly improving the computational efficiency by utilizing fast calculations such as the FFT. This also established the foundation for a novel tracking model: the correlation filter (CF).

In recent years, correlation filter-based tracking approaches enjoyed a rapid development. For precision improvement, correlation filters with multiple feature channels are designed in [82] to adopt more discriminative features such as improved HOG [50], color attributes [40] and intensity channels [48]. In [71], Henriques et al. explore essential properties of a data matrix consisting of circular shifted samples and prove that these properties are also valid for specific kernels. By adopting the kernel function, both the precision and runtime performance of their tracker have been further improved. To enhance the classifier training, several strategies are proposed. For example, spatial regularization is adopted in [38, 83] to weight samples with respect to their distance to the target. Spatial priors are utilized in [103] to handle occlusions. Support vectors [177] and context information [43] are employed to augment training samples. Multi-memory stores are introduced by [73] to accommodate appearance variations. In [37] the training set is adaptively managed to suppress the influence by imperfect or noised samples. Some works also leverage deep learning techniques by incorporating powerful deep features into their models [39,98]. Due to these efforts, CF-based trackers currently achieve state-of-the-art performance on standard benchmarks. The excellence of the correlation filter in terms of both tracking precision and efficiency also inspired the development of approaches proposed in this thesis.

2.2 Learning Correlation Filter

In this section, we introduce the details about the correlation filter from two perspectives: machine learning and signal processing. In the first perspective, we describe how to apply a classifier (which is referred to the SVM here) in the object tracking task. In the second perspective, we give a discussion about the special data structure utilized by the correlation filter: the circulant matrix, which enables a fast calculation in the frequency domain.

2.2.1 Tracking by Support Vector Machine

Machine learning is the key technique to discriminative tracking approaches, which is utilized to learn a classifier to distinguish the target from its background. Given image samples assigned with target tags at the initial frame (which is mostly the case for single-object tracking), the discriminative tracker can be learned in a supervised fashion.

Generally speaking, the supervised learning can be expressed as learning a real-valued function (a.k.a. classification/evaluation function) $f : X \subseteq \mathbb{R}^m \to \mathbb{R}$, which maps a set of samples (i.e., inputs) $\mathbf{x}_i \in X$ to their assigned labels (i.e., outputs/responses) $y_i \in \mathbb{R}$[1], where index i is in the range of $1 \leq i \leq n$ and n is the total sample number. Since samples are usually represented in the form of

[1] For classification tasks, the labels are integers, i.e., $y_i \in \mathbb{N}$. However, the Correlation Filter introduced here is based on the regression form, which takes real-valued labels. Thus, to be consistent with the description of CF and also for a more generalized interpretation, here it is written in the form of $y_i \in \mathbb{R}$.

image patches or attribute/feature vectors, the dimension m indicates the pixel number or vector length.

Support Vector Machine for Classification

In terms of conventional image classification, we usually deal with binary classification problems, i.e., to distinguish one specific object (or object class) from other objects. In this sense, the labels can only be assigned with two different values, e.g., with $y_i = 1$ for positive samples and $y_i = -1$ for negative ones. A simple idea to distinguish these samples is to construct a hyperplane (a.k.a. classification plane) to divide the space of X into two parts, each associated with a unique class label, as illustrated in Fig. 2.1 (a). This concept yields the linear form of support vector machine, in which the hyperplane is interpreted by the linear function

$$f(\mathbf{x}) = \mathbf{w}^{\top} \cdot \mathbf{x} + b, \tag{2.1}$$

where vector $\mathbf{w}$ encodes the plane parameter (i.e., the norm vector) and term b is the bias. For a more generalized representation, we reformulate Eq. (2.1) in a homogeneous form[2] which omits the bias term b, expressed as

$$f(\mathbf{x}') = \begin{bmatrix} \mathbf{w}^{\top} & b \end{bmatrix} \cdot \begin{bmatrix} \mathbf{x} \\ 1 \end{bmatrix} = \mathbf{w}'^{\top} \cdot \mathbf{x}', \tag{2.2}$$

where $\mathbf{w}'$ denotes the extended parameter vector and $\mathbf{x}'$ is the extended sample attribute. Thus, in this new form, the bias term b is included in the parameter vector describing the hyperplane in the extended feature space. This new formulation can also be considered as moving the hyperplane described in Eq. (2.1) to make it cross the zero-point of the space. The same translation is also applied on all sample points. Since the classification result is not changed,

[2] In some literatures [30, 135], the bias term b is explicitly handled as in Eq. (2.1) and not regularized during optimization. However, recent studies [47, 136] show that by utilizing the extended parameter vector $\mathbf{w}'$ (i.e., implicitly regularizing b in Eq. (2.3)), a much faster convergence rate in the learning procedure can be achieved due to a stronger convexity of the loss function. It is also found that the influence on classification performance is only limited by regularizing the bias term b. For the same reason, the form (2.2) is preferred in the work [70], which introduced the KCF tracker. For consistency, such form is utilized in this thesis.

for a simplified representation but without causing confusion, we use symbol $\mathbf{x}$ instead of $\mathbf{x}'$ and $\mathbf{w}$ instead of $\mathbf{w}'$ in the following discussion.

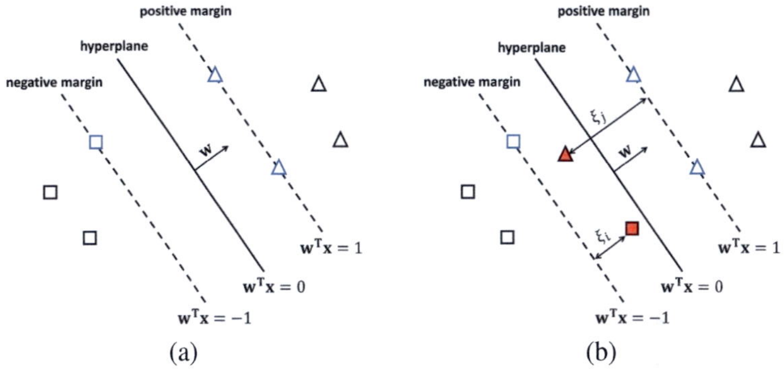

Figure 2.1: Linear support vector machine for binary classification: (a) with hard margin, (b) with soft margin. Positive samples are represented by triangles while negative samples are represented by squares. Hard samples are denoted in red. Support vectors are marked in blue. The hard margin plane is represented by dash line. The normal vector $\mathbf{w}$ of the hyperplane is represented by an arrow.

The goal of a learning approach is to approximate the true labels (a.k.a. ground truths) with the function f as accurate as possible. Thus, the output value $f(\mathbf{x})$ should also be binarized, e.g., by taking its sign. Hence, it desires $f(\mathbf{x}) \geq 0$ for positive samples and $f(\mathbf{x}) < 0$ for negative ones. Normally, we also expect that the learned classifier can perform well on untrained data. Therefore, the most stable hyperplane should guarantee the largest distance/margin to each class, which is determined by its nearest sample points (i.e., support vectors). Mathematically, it can be interpreted as solving problem

$$\arg\min_{\mathbf{w}} \|\mathbf{w}\|^2 + \frac{1}{\lambda}\sum_{i=1}^{n} \xi_i \tag{2.3}$$

subject to

$$\begin{cases} y_i \mathbf{w}^\top \mathbf{x}_i \geq 1 - \xi_i \\ \xi_i \geq 0 \end{cases}, \quad \forall i \in \{1, \ldots, n\} \tag{2.4}$$

with the non-negative slack variable ξ_i and a positive regularization factor λ. As mentioned, the bias term b is already incorporated into the augmented parameter vector $\mathbf{w}$. According to [136], regularizing the bias during optimization only leads to limited influence of the classification performance but yields a stronger convexity of the loss function, which makes the learning procedure much easier. In constraint (2.4), if all slack variables are forced to be zero, then only the first term is left in loss function (2.3), which yields the *hard-margin binary SVM* (Fig. 2.1 (a)). Recall that the distance of a sample $\mathbf{x}_i$ with respect to the hyperplane equals $\mathbf{w}^\top \mathbf{x}_i / \|\mathbf{w}\|$, the resulted constraint (2.4), i.e., $y_i \mathbf{w}^\top \mathbf{x_i} \geq 1$, forces all samples to be located on the correct side of the hyperplane with a minimum distance of $1/\|\mathbf{w}\|$, which is maximized by minimizing its denominator.

However, in most of classification tasks, we cannot find a hyperplane which can clearly separate different object classes, due to noises in samples or insufficient feature representation. Therefore, we employ a positive slack variable ξ_i in constraint (2.4) for each misclassified sample. This yields the *soft-margin binary SVM*, which tolerates samples located on the wrong side of the hyperplane. Thus, the second term in problem (2.3) is to minimize the summed distance of misclassified samples with respect to their corresponding margin planes (illustrated in Fig. 2.1 (b)). The regularization factor is just to balance between both terms that are to be minimized.

Support Vector Machine for Tracking

Although the binary SVM is utilized for most classification tasks, there are still issues restricting its application for tracking approaches. The first one is that solving the problem (2.3) should leverage the Quadratic Programming (QP) or Lagrange Multiplier Method. The resulted computation amount is high, which is inappropriate for learning trackers with strict runtime requirements. Another issue is that the location of the target should also be identified in a tracking task. This requires ranking of locations, where the classifier is evaluated, according to their scores. Thus, the SVM should be able to output continuous values instead of binary ones.

To guarantee a precise localization, ideally, the highest classification score should be obtained at the location where the detection window exactly covers the target and both of them are center-aligned. This requires that in the training process positive samples, e.g., extracted from different locations, should be assigned with non-equal labels, with low values to penalize samples with great location errors. A common approach is that the sample label is determined by a predefined probability function, e.g., a 2D Gaussian function, with respect to the distance between the center of the sample and the original target location, as illustrated in Fig. 2.2. Thus, the greatest label is always assigned to the sample located at the target center and equals to one while the smallest label is assigned to samples without overlap with the target and is set to nearly zero.

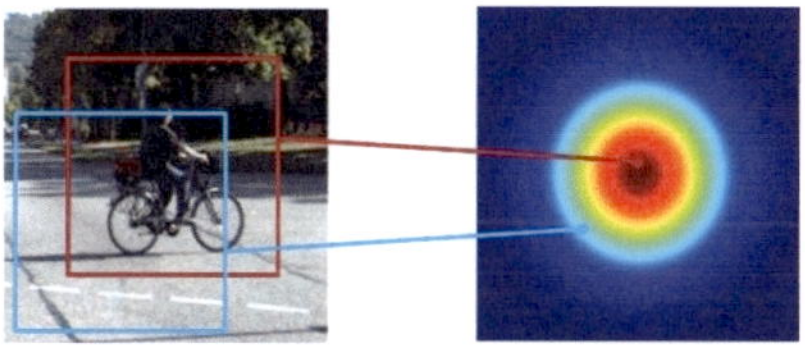

Figure 2.2: Continuous label values for samples. On the left is an image (from the KITTI dataset [95]) with the target, i.e., a cyclist, located in its center. Two samples covering the target are extracted by a smaller bounding box, respectively denoted in red and blue. On the right, it shows the label map calculated by a predefined 2D Gaussian function with respect to the distance between the sample center and the original target location. The labels are displayed by a heatmap with warm colors to indicate high values. The corresponding points of those two samples are also denoted in this map. The utilized bounding box is bigger than the target size, because we would like to include texture information from background, which is helpful for training the classifier.

Since the output label becomes a continuous real number, the objective function for binary classification is ineligible. Here we resort to the regression form of support vector machine, e.g., the linear epsilon-insensitive SVM (ϵ - SVM), which is expressed as

$$\arg\min_{\mathbf{w}} \|\mathbf{w}\|^2 + \frac{1}{\lambda}\sum_{i=1}^{n}(\xi_i + \xi_i^*) \tag{2.5}$$

subject to

$$\begin{cases} y_i - \mathbf{w}^\top \mathbf{x}_i \leq \epsilon + \xi_i \\ \mathbf{w}^\top \mathbf{x}_i - y_i \leq \epsilon + \xi_i^* \\ \xi_i \geq 0 \\ \xi_i^* \geq 0 \end{cases}, \forall i \in \{1, \ldots, n\}, \tag{2.6}$$

where ϵ controls the tolerance band for the prediction and term ξ_i and ξ_i^* are slack variables. Similar as in the soft-margin binary SVM, the slack variables represent the distance of outliers to the margin of the tolerance band. To simplify the representation, we rewrite the problem (2.5) as

$$\arg\min_{\mathbf{w}} \sum_{i=1}^{n} L_\epsilon(y_i, f(\mathbf{x}_i, \mathbf{w})) + \lambda \|\mathbf{w}\|^2 \tag{2.7}$$

subject to

$$L_\epsilon(y_i, f(\mathbf{x}_i, \mathbf{w})) = \begin{cases} 0, & \text{for } |y_i - f(\mathbf{x}_i, \mathbf{w})| \leq \epsilon \\ |y_i - f(\mathbf{x}_i, \mathbf{w})| - \epsilon, & \text{otherwise} \end{cases} \tag{2.8}$$

and the constraint can be further replaced by a hinge loss function, which is expressed as

$$L_\epsilon(y_i, f(\mathbf{x}_i, \mathbf{w})) = \max\{0, |y_i - f(\mathbf{x}_i, \mathbf{w})| - \epsilon\}. \tag{2.9}$$

Intuitively, the optimization problem of (2.7) turns to be more complicated than a binary SVM and an analytical solution is difficult to attain due to the non-differentiable form (i.e., the L1-form) of constraint in (2.9).

To circumvent this problem, we employ an alternative loss function with squared prediction errors, yielding the L2-form SVM [135], formulated as

$$\arg\min_{\mathbf{w}} \sum_{i=1}^{n} |y_i - f(\mathbf{x}_i, \mathbf{w})|^2 + \lambda \|\mathbf{w}\|^2. \tag{2.10}$$

Here we force the tolerance band ϵ to be zero to further simplify the solving procedure. Compared with the L1-form SVM, the L2-form SVM (a.k.a. L2-SVM) is differentiable and strongly penalizes samples with bigger prediction errors. Thus, it usually yields a better performance [87, 143].

For a compact representation, we leverage the linear form of function $f(\mathbf{x}, \mathbf{w})$ and rewrite the problem (2.10) as

$$\arg\min_{\mathbf{w}} \|\mathbf{y} - \mathbf{X}\mathbf{w}\|^2 + \lambda \|\mathbf{w}\|^2 \tag{2.11}$$

with label vector $\mathbf{y} = [y_1, \ldots, y_n]^\top$ and data matrix $\mathbf{X} = [\mathbf{x}_1, \ldots, \mathbf{x}_n]^\top$. Another merit of the quadratic form of the L2-SVM is that its Hessian matrix is always positive definite [87], which makes the optimization problem of (2.11) convex and thus a closed-form solution feasible. Here we summarize all to be penalized terms from Eq. (2.11) in one loss function as

$$L(\mathbf{w}) = \frac{1}{2}(\|\mathbf{y} - \mathbf{X}\mathbf{w}\|^2 + \lambda\|\mathbf{w}\|^2), \tag{2.12}$$

where the constant factor $\frac{1}{2}$ has no influence on the solution. According to its convexity property, the minimum loss value should be located at the point with zero gradients. Thus, we take the derivatives with respect to vector $\mathbf{w}$ and set them to zero, which yields

$$\frac{\partial L(\mathbf{w})}{\partial \mathbf{w}} = -\mathbf{X}^\top\mathbf{y} + (\mathbf{X}^\top\mathbf{X} + \lambda\mathbf{I})\mathbf{w} = \mathbf{0}, \tag{2.13}$$

where term $\mathbf{I}$ and $\mathbf{0}$ respectively denote the identity matrix and zero vector. By solving Eq. (2.13), we obtain the analytical solution of problem (2.11), interpreted as

$$\mathbf{w} = (\mathbf{X}^\top\mathbf{X} + \lambda\mathbf{I})^{-1}\mathbf{X}^\top\mathbf{y}. \tag{2.14}$$

The form of Eq. (2.14) is also known as the *ridge regression*. Compared with the *least square estimation*, it employs an additional regularization term λ to deal with cases where the matrix $\mathbf{X}^\top\mathbf{X}$ is ill-conditioned [30].

2.2.2 Circular Data Structure

Although the closed-form solution of L2-SVM is not complicated, the explicit inversion of matrix in Eq. (2.14) is still the bottleneck for fast computation, especially when the data matrix $\mathbf{X}$ consists of a large number of samples. Moreover, the time cost of the sampling approach is proportional to the number of samples, which are usually extracted from different image locations. Since a big training dataset is commonly considered as the key to a good generalization ability, in this sense, the contradiction between the classification power of a SVM and its runtime performance (including learning) seems irreconcilable.

Since preparing the training dataset could be laborious and computationally expensive, recent research works [44, 116] resort to learning with virtual samples which yields a performance comparable with that trained on real images. The success of their idea led us to the belief that our problem could also be solved in a similar way. Considering that in a tracking task training samples are densely extracted from a small nearby region of the target, most of the samples overlap with each other and thus are strongly correlated. Such relationship can be equivalently interpreted by a translation operation, i.e., circular shifting, to transform the original target image patch. For a small shifting step, the transformed image sample still shares a big area with the original one and only differs in a few pixels at the margin (Fig. 2.3 (b)). For a big translation, the artifact caused by circular shifting becomes severe (Fig. 2.3 (e)). However, as the transformed sample is assigned with a very small label value, i.e., nearly zero, it will be taken as a negative sample and thus imposes very limited influence on learning the target appearance. In this way, the circular shifting approach merely changes the prior knowledge of the target appearance. Moreover, since only one image sample (which contains the target) is required, the preciously mentioned explicit sampling procedure can be spared while the amount of training data remains unchanged. Furthermore, the circular shifting yields a very special form of the data matrix $\mathbf{X}$, i.e., the circulant matrix[3] [74], which significantly eases the calculation of Eq. (2.14).

(a) (b) (c) (d) (e)

Figure 2.3: Circular shifting in horizontal direction. On the leftmost side is the original image sample containing the target, i.e., a cyclist. The other four samples are its circular shifted versions with a step size of 13 pixels. Image sample is from the KITTI dataset [95].

Circulant Matrix

Principally, the circular representation can be generalized for 2D images and multi-channel feature maps [70]. However, for a simplified representation of the circulant matrix, we only consider the one-dimensional case, where the

[3] In literatures, the circulant matrix is usually referred to the transposed matrix $\mathbf{X}^\top$. However, for simplified representation, we directly derive the properties introduced in this thesis from the matrix $\mathbf{X}$ rather than from its transposed form.

target sample is interpreted as a vector[4] $\mathbf{x} = [x_1, \ldots, x_n]^\top$. Considering the shifting step equal to 1, the data matrix $\mathbf{X}$ can be written as

$$\mathbf{X} = \begin{bmatrix} x_1 & x_2 & \cdots & x_{n-1} & x_n \\ x_n & x_1 & \cdots & x_{n-2} & x_{n-1} \\ \vdots & \vdots & \ddots & \vdots & \vdots \\ x_3 & x_4 & \cdots & x_1 & x_2 \\ x_2 & x_3 & \cdots & x_n & x_1 \end{bmatrix} = \begin{bmatrix} (\mathbf{P}^0\mathbf{x})^\top \\ (\mathbf{P}^1\mathbf{x})^\top \\ \vdots \\ (\mathbf{P}^{n-2}\mathbf{x})^\top \\ (\mathbf{P}^{n-1}\mathbf{x})^\top \end{bmatrix}, \tag{2.15}$$

where $\mathbf{P}$ is a permutation matrix with

$$\mathbf{P}^i\mathbf{x} = [x_{n-i+1}, \ldots, x_n, x_1, \ldots, x_{n-i}]^\top \quad \text{for} \quad 1 \leq i \leq n. \tag{2.16}$$

From the above formulation, we can see two interesting points: Firstly, since each row of matrix $\mathbf{X}$ consists of one sample, the total row number is equal to the dimension of $\mathbf{x}$. Hence, the amount of training data solely depends on the resolution of the sample image or its feature map. Secondly, the circular shifting operation can be cast into a multiplication between sample $\mathbf{x}$ and a permutation matrix $\mathbf{P}$, which can be efficiently implemented on modern hardwares such as digital signal processors (DSPs).

Recall that the Discrete Fourier Transform (DFT) also adopts a complex multiplier with periodic powers [34], a link between the circulant matrix and the DFT can be easily established [41], interpreted as

$$\mathbf{X} = \mathbf{F}^H \operatorname{diag}(\hat{\mathbf{x}})\mathbf{F}, \tag{2.17}$$

where term $\operatorname{diag}(\hat{\mathbf{x}})$ is a diagonal matrix. Its diagonal is formed by vector $\hat{\mathbf{x}}$, which is the DFT form of sample $\mathbf{x}$ and can be expressed as $\hat{\mathbf{x}} = \mathcal{F}(\mathbf{x})$ with the Fourier operator $\mathcal{F}$. The matrix $\mathbf{F}$ is called as *DFT matrix*, which consists of constant complex numbers and reformulates the DFT as a matrix multiplication: $\mathcal{F}(\mathbf{x}) = \sqrt{n}\,\mathbf{F}\mathbf{x}$. According to this formulation, the matrix $\mathbf{F}$ is unitary,

[4] Previously, the feature dimension is defined as m and the sample number is n. However, for a circulant matrix generated by one-dimensional vector, both values are equal.

which means $\mathbf{F}^H\mathbf{F} = \mathbf{I}$ with superscript ^{H} to indicate the Hermitian transpose. Additionally, for each vector $\mathbf{x}$, we have the norm-preserving property of

$$||\mathbf{F}\mathbf{x}||^2 = ||\mathbf{x}||^2 = ||\mathbf{F}^H\mathbf{x}||^2. \tag{2.18}$$

Fast Calculation in Frequency Domain

Enlightened by the norm-preserving property (2.18), the loss function (2.11) will not be changed if the unitary matrix $\mathbf{F}$ is multiplied within each of its terms. Thus, we can rewrite it as

$$L(\mathbf{w}) = \frac{1}{2}(||\mathbf{F}\mathbf{y} - \mathbf{F}\mathbf{X}\mathbf{w}||^2 + \lambda||\mathbf{F}\mathbf{w}||^2). \tag{2.19}$$

Leveraging the relationship (2.17) and the unitary property of matrix $\mathbf{F}$, loss function $L(\mathbf{w})$ can be further reinterpreted as

$$L(\mathbf{w}) = \frac{1}{2}(||\mathbf{F}\mathbf{y} - \mathrm{diag}(\hat{\mathbf{x}})\mathbf{F}\mathbf{w}||^2 + \lambda||\mathbf{F}\mathbf{w}||^2). \tag{2.20}$$

Based on the definition of DFT matrix, all vectors multiplied with $\mathbf{F}$ in the above form can be replaced with their Fourier transforms. Hence, we obtain a new loss function

$$L(\hat{\mathbf{w}}) = \frac{1}{2n}(||\hat{\mathbf{y}} - \mathrm{diag}(\hat{\mathbf{x}})\hat{\mathbf{w}}||^2 + \lambda||\hat{\mathbf{w}}||^2) \tag{2.21}$$

which is only represented by terms from the frequency domain. The constant multiplier $\frac{1}{2n}$ can be easily omitted, because it imposes no influence on the optimization results. Compared with the original loss function (2.11), the computation amount in this new form is low due to the reason that in the product $\mathrm{diag}(\hat{\mathbf{x}})\hat{\mathbf{w}}$, non-zero multiplication can only happen on diagonal elements.

For a better understanding of the new form (2.21), we reformulate the matrix product $\mathrm{diag}(\hat{\mathbf{x}})\hat{\mathbf{w}}$ as

$$\mathrm{diag}(\hat{\mathbf{x}})\hat{\mathbf{w}} = \hat{\mathbf{x}} \odot \hat{\mathbf{w}} = \mathcal{F}(\mathbf{x} * \mathbf{w}) = \mathcal{F}(\mathbf{X}\mathbf{w}), \tag{2.22}$$

where $\odot$ indicates the Hadamard product and operator $*$ denotes the convolution. Eq. (2.22) points out that since matrix $\mathbf{X}$ only consists of circular shifted versions of sample $\mathbf{x}$, the matrix multiplication $\mathbf{Xw}$ can be considered as the convolution between a filter $\mathbf{w}$ and image $\mathbf{x}$. According to signal processing theory, the convolution in spatial domain can be replaced by the element-wise product between their Fourier transforms. In this sense, solving the original loss function $L(\mathbf{w})$ is equivalent to solve its Fourier form $L(\hat{\mathbf{w}})$ in frequency domain, which is consistent with the Parseval's theorem.

In analogy to its form in spatial domain, the optimum of $L(\hat{\mathbf{w}})$ is also located at its zero-gradient point. Thus, by setting its derivatives to zero and solving those equations, the solution is obtained as

$$\hat{\mathbf{w}} = \frac{\hat{\mathbf{x}}^{\star} \odot \hat{\mathbf{y}}}{\hat{\mathbf{x}}^{\star} \odot \hat{\mathbf{x}} + \lambda}, \tag{2.23}$$

where superscript * indicates the conjugate of a complex number and the division is conducted element-wise. The spatial form $\mathbf{w}$ can be easily obtained by an inverse Fourier transform expressed as $\mathbf{w} = \mathcal{F}^{-1}(\hat{\mathbf{w}})$. Compared with the original solution form (2.14), the explicit matrix inversion disappears and only element-wise operations are conducted in the new form (2.23). The computational overhead by additional Fourier transform for these utilized vectors is also little, e.g., by deploying the method of *Fast Fourier Transform (FFT)*. The computational complexity for the new solution form is only $O(n \log n)$ while for the original form it is normally $O(n^3)$ (reported on state-of-the-art solvers such as [27]).

To recognize the target as well as to reuse the fast calculation fashion, for a new image sample $\mathbf{z}$, we apply the evaluation function on all of its circularly shifted versions (contained in matrix $\mathbf{Z}$), formulated as

$$\mathbf{f}(\mathbf{z}) = \mathbf{Zw} = \mathbf{z} * \mathbf{w} = \mathcal{F}^{-1}(\hat{\mathbf{z}} \odot \hat{\mathbf{w}}). \tag{2.24}$$

The point with the maximum value is then searched in the response map $\mathbf{f}(\mathbf{z})$[5] and considered as the location of the target. In this way, both classification[6]

[5] Here we distinguish between two annotations $\mathbf{f}(\mathbf{z})$ and $f(\mathbf{z})$, where the first one indicates the evaluation on all shifted versions of sample $\mathbf{z}$ and thus it represents the response map, while the second one indicates the single evaluation on sample $\mathbf{z}$ and thus it is scalar.

[6] Here we consider the utilized regression form of SVM as a general form for classification with real-valued labels.

and localization tasks are joined in one step. Since the evaluation is performed in a correlation-like manner (which resembles the convolution operation yet with a rotated filter), such discriminative tracking approach is named as the correlation filter.

In above discussions, we only considered sample $\mathbf{z}$ as a single-channel vector. However, in most classification and tracking tasks, we usually use multiple feature channels encoding different information resources as representation attributes. Since these channels are constructed independently from each other, evaluation function (2.24) can be easily extended to multiple channel features. Thus, given a total channel number C, the evaluation on all shifted versions of sample $\mathbf{z}$ can be expressed as

$$\mathbf{f}(\mathbf{z}) = \sum_{c=1}^{C} \mathbf{z}_c * \mathbf{w}_c = \mathcal{F}^{-1}\left(\sum_{c=1}^{C} \hat{\mathbf{z}}_c \odot \hat{\mathbf{w}}_c\right), \tag{2.25}$$

where term $\mathbf{z}_c$ denotes the sample features from channel c and $\mathbf{w}_c$ contains the corresponding filter coefficients for current channel, which is learned in the same form of Eq. (2.23) only with a minor modification by replacing sample $\hat{\mathbf{x}}$ with one of its feature channel $\hat{\mathbf{x}}_c$.

2.3 Specialized Correlation Filter

The correlation filter has been widely applied in current tracking approaches. In this section we introduce one of the specific strategies to further improve the learning power of the correlation filter, which is also adopted in the approaches proposed in this thesis.

Kernelized Correlation Filter: Basic Theory

Although the linear SVM can quickly find a model to fit the expected outputs, its approximation precision may still be less satisfied for complex datasets, in which lots of samples are located far from the estimated hyperplane. Such a problem is usually caused by the fact that the exploited attribute/feature representation is insufficient to model the distribution of samples. Thus, the representation of samples in a better feature space, e.g., with higher dimensions, is greatly desired.

With this point in mind, we define a new mapping function $\phi : \mathbb{R}^m \rightarrow \mathbb{R}^{m''}$, which transforms the original sample attributes into a high dimensional space with $m'' \geq m$, interpreted as $\mathbf{x}'' = \phi(\mathbf{x})$, where $\mathbf{x}''$ is the sample representation in the new feature space. According to the Representer Theorem [134], the solution $\mathbf{w}$ for a linear regression problem belongs to the space spanned by its samples, which is interpreted as

$$\mathbf{w} = \sum_{i=1}^{n} \alpha_i \phi(\mathbf{x}_i) = \mathbf{X}'' \alpha \tag{2.26}$$

with coefficient vector $\boldsymbol{\alpha} = [\alpha_1, \ldots, \alpha_n]^\top$ and matrix $\mathbf{X}''$ consists of all transformed samples with each in one row. Hence, the evaluation on mapped attributes of an image sample $\mathbf{x}$ can be reformulated as

$$f(\mathbf{x}) = \mathbf{w}^\top \phi(\mathbf{x}) = \sum_{i=1}^{n} \alpha_i \phi(\mathbf{x}_i)^\top \phi(\mathbf{x}) = \sum_{i=1}^{n} \alpha_i \kappa(\mathbf{x}_i, \mathbf{x}), \tag{2.27}$$

where κ denotes a kernel function which implicitly calculates the dot-product $\phi(\mathbf{x}_i)^\top \phi(\mathbf{x})$. Intuitively, the calculation of $f(\mathbf{x})$ becomes more difficult, especially when ϕ is a very complex mapping function. However, if we observe Eq. (2.27) carefully, we can find out that the evaluation result sorely depends on the utilized kernel function κ. This inspires us to the idea that we can build the kernel κ with simple functions (e.g., the Gaussian) to indirectly evaluate the product $\phi(\mathbf{x}_i)^\top \phi(\mathbf{x})$ to circumvent the explicit transform of samples. Introducing such concept into the normal classification problem yields the new definition: *kernelized support vector machine*. Analogously, the type of CF tracker which employs kernel function is called as *kernelized correlation filter (kernelized CF or KCF)* [71].

To learn the KCF is just to solve the coefficient vector α by minimizing the loss function (2.11). Since the vector α is directly related to filter $\mathbf{w}$, we just introduce Eq. (2.27) into the original solution (2.14) and obtain

$$\alpha = (\mathbf{K} + \lambda \mathbf{I})^{-1} \mathbf{y}, \tag{2.28}$$

where $\mathbf{K}$ is the kernel matrix with each element equal to $k_{ij} = \kappa(\mathbf{x}_i, \mathbf{x}_j)$. Leveraging the relationship between circulant matrix and the Fourier transform, we introduce Eq. (2.17) into the above solution and thus obtain the Fourier transform of vector $\boldsymbol{\alpha}$, interpreted as

$$\hat{\alpha} = \frac{\hat{\mathbf{y}}}{\hat{\mathbf{k}}^{\mathbf{xx}} + \lambda}, \tag{2.29}$$

where the division is conducted element-wise and the vector $\hat{\mathbf{k}}^{\mathbf{xx}}$ is the Fourier form of the first row of matrix $\mathbf{K}$.

In a similar manner, by introducing Eq. (2.27) into Eq. (2.24), the evaluation of KCF on an image sample $\mathbf{z}$ as well as on all of its circularly shifted versions can be written as

$$\mathbf{f}(\mathbf{z}) = \mathcal{F}^{-1}(\hat{\mathbf{k}}^{\mathbf{xz}} \odot \hat{\alpha}), \tag{2.30}$$

where $\hat{\mathbf{k}}^{\mathbf{xz}}$ can be considered as the Fourier form of kernelized convolution[7] between training sample $\mathbf{x}$ and test sample $\mathbf{z}$.

To cope with the appearance variation of the target, the appearance template $\mathbf{x}$ and coefficient α is usually updated in an interpolation manner with a small positive learning rate β, interpreted as

$$\begin{cases} \mathbf{x}_l = (1-\beta)\mathbf{x}_{l-1} + \beta\mathbf{x}_{cur} & (2.31a) \\ \alpha_l = (1-\beta)\alpha_{l-1} + \beta\alpha_{cur} & (2.31b) \end{cases},$$

where vector $\mathbf{x}_l$ and $\mathbf{x}_{l-1}$ respectively denote the appearance attributes stored at frame l and $l-1$, while $\mathbf{x}_{cur}$ stands for the one learned from the current image. Similar subscript notations are also utilized for coefficient vector $\boldsymbol{\alpha}$.

Kernelized Correlation Filter: Merits and Shortcomings

As introduced above, similar to linear correlation filters, both the learning and inference procedure of KCF can be accomplished in the frequency domain. Since only element-wise operations are conducted and fast calculation methods

[7] In some literatures, it is also called as kernelized cross-correlation, since the correlation operation can be rewritten as convolution by rotating the filter by 180°.

such as FFT can be utilized, the computational complexity of KCF is little[8], which is also $O(n \log n)$. As the samples are mapped into a feature space with higher dimensions, the estimated hyperplane by KCF enjoys a better fitting of the desired outputs. Thus, it can yield a superior performance over conventional linear correlation filters and other conventional tracking approaches such as optical flow, which is demonstrated in [71].

However, due to the design concept, there are still challenging scenarios where the KCF performs imperfectly. For instance, the KCF is not explicitly designed to recognize occluded targets. Especially in cases with severe occlusions, the target suffers from great appearance change, which may lead to detection failures. Besides, in a lot of kernel functions, e.g., the Gaussian kernel, multiple feature channels of a sample are treated equally (shown in following chapters), which is inappropriate in scenarios such as at night, where visual features suffer from different degradations. Moreover, the linear updating fashion of KCF does not reject corrupted training samples (caused by deteriorated vision condition such as the raindrop), which may lead to tracking drifts or even failures. Furthermore, the KCF is designed mainly for tracking a single target. For the task of multiple target reidentification, additional association strategies are imperatively required.

In the following chapters, we will introduce the novel tracking approaches proposed in this thesis, which are based on KCF but aimed to tackle each of above-mentioned problems.

[8] For most utilized kernel functions, e.g., the Gaussian kernel, they can be efficiently calculated in frequency domain, which only yields little computational overhead. This will be shown in the following chapters.

3 Tracking with Severe Occlusion

Despite tremendous progress achieved by recent trackers [138], the task of visual tracking is still challenging, especially in dealing with severe occlusions, where the visible part of an object is very limited, which leads to negative impact on both the learning and inference procedure of tracker model and further results in tracking failures. Such case is common in traffic scenarios, especially for the VRUs, i.e., the pedestrians and bicyclists. Because of a relative small size, the VRUs can be easily occluded by big objects like vehicles on the road and only small body parts are left noticeable [146, 151], as shown in Fig. 3.1. This is a typical scenario in every day life (e.g., at narrow streets or before road intersections) but a very difficult one to deal with.

Conventional vision-based frameworks such as [149, 150] regularly fail to perceive and track those partially occluded objects because of four main reasons [146, 151]: Firstly, the unexpected appearance change from full body to small visible object parts can result in matching failures. Secondly, to perceive these small obvious parts, it requires extra detection process, which may end up troublesome when the obvious parts are not so distinctive. Thirdly, if long term occlusion shows up, which prompts long time loss of tracked object, the appearance model can be debased by a plenty of false positives and cannot recuperate. Last but not least, despite that motion information can be utilized, the object location can still be misestimated, if object's motion unexpectedly changes (e.g., by sudden speeding up or brake).

Aiming to tackle the above problems, in this chapter, we present a novel approach to track objects under severe occlusion by utilizing part filters, which is published in works [146, 151]. The organization of this chapter is as follows: Firstly, we introduce the related tracking approaches on dealing with occlusions. Afterwards, we give a description about the proposed mechanism for occlusion awareness. This part includes two main points: occlusion occurrence detection and occluded object parts identification. Thereafter, a dynamic framework is presented to efficiently manage utilized filters. Finally, evaluation results of the proposed method on several datasets are given, demonstrating that it performs superior over state-of-the-art especially in dealing with long term severe occlusions.

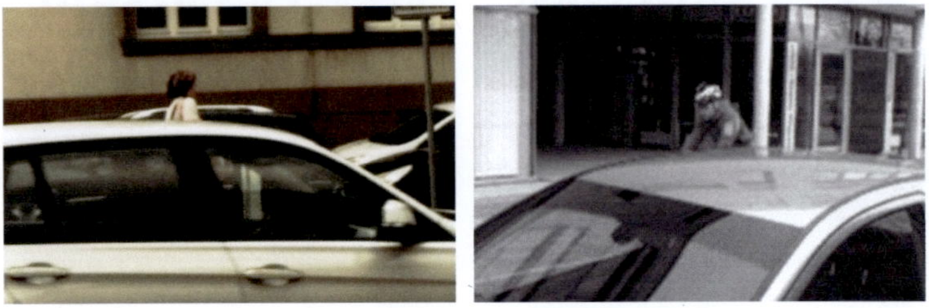

(a) (b)

Figure 3.1: In image (a) and (b), a pedestrian and a cyclist are respectively occluded by vehicles on the road [151]. Just their heads and shoulders are noticeable in the image.

To avoid confusion, the term occlusion addressed in this chapter, without explicit statement, refers to the partial occlusion. As to full occlusions (particularly the long term ones), since the target completely vanishes in the image, it cannot be recognized any more. Once the target is rediscovered, e.g., after a long time full occlusion, it should be associated with its previous trajectories. This is yet based on additional association strategies, which will be addressed in following chapters. Aside from that, the terms part tracker and part filter are not differentiated in this chapter, since most part trackers, discussed in this chapter, are constructed in a filter-like structure.

3.1 State of the Art

Dealing with occlusion is considered as one of the persistent challenging issues confronted in the visual tracking community. In recent researches, this topic gains increased interest. Depending on the manner how the occlusion is handled, visual tracking approaches can be generally categorized into two subgroups: implicit and explicit approaches.

Implicit Occlusion Handling

The tracking approaches with implicit occlusion handling neither employ specific mechanisms to recognize the occurrence of occlusion nor deploy additional strategies for modeling the visible object parts. They usually adopt the same model for object tracking in both occluded and non-occluded cases. The focus of these approaches is to enhance the learning power of the classification model to make it capable to capture the most discriminative information of the target. Although there is no explicit occlusion handling in these approaches, they still report some robustness against occlusion according to their evaluation results. In the following part, we give a brief overview about the corresponding research directions as well as an introduction of some representative approaches.

In the first research direction, tracking approaches attempt to involve more background image information into the training data so that the learned classifier can better distinguish the target from other objects such as the obstacles. Such approach type usually relies on linear correlation filters. For instance, Danelljan et al. in [38] expand the input sample by a large background image margin in its surroundings. Although the diversity of samples is increased, background noises covered by the filter coefficients will also be learned by the classifier. To address this problem, they replace the regularization factor λ of loss function (2.11) by a matrix with big values to suppress the filter coefficients covering background image areas. Although their tracker is proven with improved robustness, the circular data structure is lost, which makes the fast calculation impossible. As a circumvention, Lukežič et al. in [103] directly utilize a binary mask to constrain the number of filter coefficients so that only a small image region can be evaluated. Such approach imposes no harm on the circular data structure. However, the binary mask is obtained by a segmentation approach. It may yield a very small segment due to occlusions, which further leads to a very small filter, losing the information of full target.

In another branch, research works are more interested in learning classifier from different historical samples. For instance, Hong et al. in [73] sort detected target samples in a temporal order and save them in two memory stores. In the short-term store, they train a KCF tracker based on recently collected samples while in the long-term store they learn a optical flow tracker across a long frame sequence. The inference is made by the tracker with the higher estimated reliability. In this approach, they are able to detect the target with non-severe occlusion by the long-term tracker, because the appearance model of short-term tracker can be severely contaminated when occluded samples are included in the training data while the long-term tracker is merely influenced as it emphasizes more on the old samples. In a similar concept, Nam et al. in [114] build the tracker with two separate CNNs. They adjust the updating rate so that one CNN can quickly learn the current appearance of target while the other learns the target in a long time. By manipulating multi-temporal domains, their approach achieves a boosted tracking precision yet with high computation loads.

In other directions, context information is also deployed by some researchers to assist the learning or inference of the classifier. For instance, Mueller et al. in [113] extend the loss function (2.11) to train the classifier both on the target sample and context image samples extracted from specific regions, yielding a promoted detection rate. In the work [43], context samples contain objects with similar appearance of the target. They are trained as negative samples in order to avoid confusion with the target. Additionally, key points in the background are utilized as supporters to help the inference according to their motion relationship to the target. However, the performance of this approach may be limited in some scenarios, e.g., with homogeneous backgrounds.

Explicit Occlusion Handling

According to the above discussion, we can see that tracking approaches with implicit occlusion handling already show some kind of robustness, thanks to various strategies adopted in these approaches. However, most of them are still incapable to deal with very difficult scenarios such as with long term severe occlusion. Since these approaches are mostly learned on-the-fly and updated for each frame, image samples containing the occluded target can be easily included into the training data. Due to lacking of occlusion recognition mechanism, these contaminated samples are indubitably treated as positive samples, which will misguide the learning of classification model. Even though reliability estimation is employed in some approaches, for long term severe occlusion, samples containing the correct target appearance will be unavailable for a long time. Hence, the tracker model has no chance to be updated or corrected during the occlusion. Its reliability thus degrades over the time, finally leading to tracking drifts or failures.

In contrast, approaches with explicit occlusion handling are commonly integrated with occlusion awareness mechanism and decompose the target into multiple parts with each part represented by a single model. By identifying the visible parts of the target, these approaches are more suitable in dealing with severe occlusion and usually yield a superior performance. Depending on how the part model is constructed, approaches in this category can be further sorted into two subgroups: generative part models and discriminative part models.

In the first subgroup, approaches describe each part of the target with manually designed appearance models and match them in next frame with respect to specific distance metrics or probabilistic methods. Representatively, Kwon et al. in [93] model the appearance of target by a lot of small local patches, which can be deleted or changed due to their robustness measurement. They also combine the Monte Carlo method with a local optimizer to reduce the computational complexity and achieve an efficient inference. Similarly, the target in [28] is decomposed into several segments with ellipsoidal shapes and each of them is represented by the mean and covariance of its color and spatial coordinates. The segments are matched in analogy to the Lucas-Kanade algorithm and a *Gaussian Mixture Model (GMM)* is utilized for the target inference. Although this method is able to handle occlusion, it requires the user to mark the target at the input image. In a more advanced version, Maggio et al. in [106] allow the overlap between part models and represent them by color histograms. The matching is based on the Bhattacharryya distance and their method is robust against rotation and scale variations. However, in most of these methods, the part matching operation is conducted inefficiently, thus their computational complexities are relative high.

In comparison, approaches with discriminative part models usually train individual classifiers for each part. Since efficient trackers such as the correlation filter can be deployed, their runtime performance is more favorable. For instance, Li et al. in [99] model each part of the target by a KCF tracker trained on a small patch. For each part, they estimate its reliability with respect to a specific confidence function and only exploit the most reliable ones to infer the target location by a Hough voting scheme, achieving encouraging results. In the work [1], Akin et al. employ both local and global filters respectively for tracking parts and the full object. The local filters provide an initial reference for the global filter while the global filter provides feedback to part filters regarding their deformation parameters, which are further used to estimate their reliabilities. Their method is verified with improved robustness against occlusion and deformation of objects. Different from them, Liu et al. in [100] utilize correlation filters to build part models and combine a Bayesian inference framework with structural constraints. Despite the benefit for both precision and processing time, as the part number is fixed, the question of how to set the number and the size of part models still remains open. This issue is yet crucial to deal with severely occluded objects and rarely considered in above mentioned methods.

In this chapter, a novel tracking approach is proposed based on the correlation filter to track severely occluded objects. The occlusion is recognized by bi-directional search fashion. With the help of a masking process, the visible object area can be obtained in a pixel-level precision. Additionally, both the number and size of part filters are adapted to the current target appearance.

3.2 Adaptive Part Filter Modeling

In this section, we introduce the proposed tracking approach based on adaptive part filter modeling to tackle the tracking problem with severe occlusion. Due to the high tracking precision and high computational efficiency of KCF tracker, it is chosen as the base framework for both the part filter and the full object tracker. Here a Gaussian function is used as the kernel and a brief description about it is given in the beginning of this section. In the following parts, regarding the stated issues in the beginning of this chapter, the occlusion recognition mechanism is firstly discussed, which is able to recognize the abrupt change of object appearance. Thereafter, filter construction strategies and the masking process are introduced, by which the visible object areas can be identified in a pixel-level precision. Finally, a dynamic filter management framework is presented, which can efficiently handle the updating process particularly in the case of long time occlusion. Since the proposed tracker mainly relies on the visual information, it is immune to estimation errors from motion models, which are usually employed in other tracking works.

3.2.1 KCF with Gaussian Kernel

As discussed in Chapter 2, kernel functions are adopted to map samples into a higher dimensional space to make them more distinguishable. Intuitively, the selection of kernel function κ could be discretionary. However, not all the kernels are valid for the circular data structure. According to [71], the kernel matrix $\mathbf{K}$ in Eq. (2.28) is only circulant, when the kernel function of two samples $\mathbf{x}_i$ and $\mathbf{x}_j$ satisfies

$$\kappa(\mathbf{x}_i, \mathbf{x}_j) = \kappa(\mathbf{M}\mathbf{x}_i, \mathbf{M}\mathbf{x}_j) \tag{3.1}$$

for any permutation matrix $\mathbf{M}$. The valid kernels are the ones with exponential, dot-product and additive functions. Here the Gaussian function is used as the kernel, as it is employed in most tracking works and demonstrated with a superior performance over other kernel functions.

Hence, given a predefined standard deviation σ, the utilized Gaussian kernel function can be interpreted as

$$\kappa(\mathbf{x}_i, \mathbf{x}_j) = \exp\left(-\frac{1}{\sigma^2} h(\mathbf{x}_i, \mathbf{x}_j)\right) \tag{3.2}$$

with

$$h(\mathbf{x}_i, \mathbf{x}_j) = \|\mathbf{x}_i - \mathbf{x}_j\|^2 = \|\mathbf{x}_i\|^2 + \|\mathbf{x}_j\|^2 - 2\mathbf{x}_i^\top \mathbf{x}_j. \tag{3.3}$$

Leveraging this expression, the kernelized convolution between sample $\mathbf{x}$ and $\mathbf{z}$ can be rewritten as

$$\mathbf{k}^{\mathbf{xz}} = \exp\left(-\frac{1}{\sigma^2} \mathbf{h}(\mathbf{x}, \mathbf{z})\right) \tag{3.4}$$

with

$$\mathbf{h}(\mathbf{x}, \mathbf{z}) = \|\mathbf{x}\|^2 + \|\mathbf{z}\|^2 - 2\mathbf{x} * \mathbf{z}. \tag{3.5}$$

From Eq. (3.5), it can be seen that the first two terms are both constant while the convolution term can be reinterpreted as dot-product between their Fourier transforms in the frequency domain. Thus, a fast computation is also guaranteed for the correlation filter with Gaussian kernels.

3.2.2 Occlusion Recognition Mechanism

The KCF tracker has been widely employed in tracking tasks. Despite its success in non-complex scenarios, as its model is always with fixed size and lacking of occlusion awareness, it can be difficult to track occluded objects. Considering the fact that the happening of an occlusion is ordinarily sudden, e.g., a pedestrian/cyclist occluded by a bypassing vehicle on the road, the target appearance thus experiences incredible change. This impact is further reflected in the learned appearance model of the tracker just as in its classification results. In light of this supposition, two criteria are chosen to recognize the occlusion: the peak-to-sidelobe ratio (PSR) and the normalized object difference (NOD). The PSR value estimates the sharpness of filter response and is calculated as

$$\mathrm{PSR} = \frac{\max(\mathbf{f}(\mathbf{z})) - \mu}{\sigma + \epsilon}, \tag{3.6}$$

where term μ and σ respectively denote the mean value and the standard deviation of response map obtained by applying classifier $f(\mathbf{z})$ on image sample $\mathbf{z}$. The small positive value ϵ is employed to avoid the case of division by zero. As a non-occluded target is more distinguishable than an occluded one, it is more probable to yield a sharp peak in the response map without occlusions (Fig. 3.2 (c)). In this way, the PSR value is correlated with the classification confidence and decreases if occlusion happens (Fig. 3.2 (e)). In the second criterion, it calculates differences between the previously utilized appearance model $\mathbf{x}_{pre}$ and the currently exploited $\mathbf{x}_{cur}$, which are further normalized by the object area A. The calculation can be interpreted by the following equation

$$\mathrm{NOD} = \frac{1}{A}\|\mathbf{x}_{pre} - \mathbf{x}_{cur}\|_2. \tag{3.7}$$

Thus, high values of NOD imply significant appearance change as well as high occlusion probability (Fig. 3.2 (f)). The occurrence of occlusion can thus be determined based on both above criteria, e.g., when their values are found within specified ranges, which is formulated as

$$\mathrm{PSR} < \theta_P \;\wedge\; \mathrm{NOD} > \theta_N, \tag{3.8}$$

where term θ_P and θ_N are predefined thresholds and can be learned from training datasets.

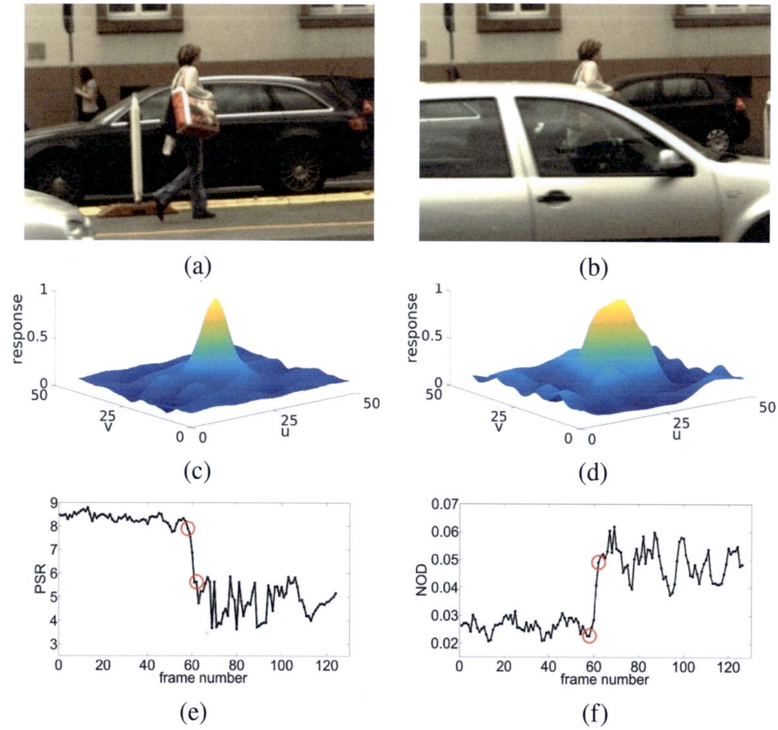

Figure 3.2: On the top row are two frames from a sequence of the driving dataset [146]. Image (a) shows a walking pedestrian at frame 58. Image (b) shows the pedestrian occluded by a vehicle at frame 62. Image (c) and (d) display the filter response of the full object tracker for both frames. The related PSR and NOD plots over the whole sequence are given in image (e) and (f) respectively while the referred frames are marked by red circles.

3.2.3 Part Filter Construction Strategy

The initial step to build part filters is to identify the visible object area. This task here is accomplished in a backward searching fashion. At first, a storage for each target is established to spare its appearance models and the corresponding images in the last q frames. For each frame, the strategy presented in Section 3.2.2 is utilized to detect the occurrence of occlusion. For a positive detection, dense feature points are initialized in the currently matched target region bounded by the detection box.

The feature points are selected from corner points in the image, which exhibit great variation of pixel value by image shifting and can be detected by the Harris-Corner-Detector [68]. Given a gray image I (e.g., converted from an RGB image) and a displacement $(\Delta x, \Delta y)$, the *sum of squared differences* of pixels within a small patch ω after shifting can be interpreted as

$$E_\Delta = \sum_{(x_i, y_i)\in\omega} (I(x_i, y_i) - I(x_i + \Delta x, y_i + \Delta y))^2, \tag{3.9}$$

where (x_i, y_i) denote the pixel coordinates. According to [68], by leveraging the Taylor expansion, above equation can be approximated as

$$E_\Delta = [\Delta x, \Delta y]\,\mathbf{M}\begin{bmatrix}\Delta x\\ \Delta y\end{bmatrix} \tag{3.10}$$

with

$$\mathbf{M} = \sum_{(x_i, y_i)\in\omega}\begin{bmatrix} I_{x,i}^2 & I_{x,i}\cdot I_{y,i}\\ I_{x,i}\cdot I_{y,i} & I_{y,i}^2\end{bmatrix}, \tag{3.11}$$

where $I_{x,i}$ and $I_{y,i}$ respectively denote the derivatives of pixel $I(x_i, y_i)$ in x and y directions. For a corner point, eigenvalues of matrix $\mathbf{M}$ should be large. In the method [68], this is equivalent to check the quality score

$$R = \det(\mathbf{M}) - \rho(\mathrm{trace}(\mathbf{M}))^2, \tag{3.12}$$

where parameter ρ is empirically set to 0.04. Local maxima which are greater than a predefined threshold θ_R (set to 15 in experiments) are considered as corner points.

In this approach, the feature points are searched in a grid fashion. The image region is firstly divided into a number of 10×10 grid cells. Then the Harris-Corner-Detector is applied to each cell to detect corner points. If multiple points are detected in one cell, only the point with the maximum score is kept. Afterwards, the non-maximum-suppression operation with a window of 8×8 pixels is applied over grid cells to remove closely located points. Furthermore, points falling into a predefined margin to box boundaries are also removed, so that the number of points selected from the background can be significantly reduced while the points selected from the central area of the box can remain. This is consistent with the fact that the target is mostly located at the box center. Thereafter, correspondences are searched for these points in each saved frame based on an optical flow approach [102]. Here only the points which are shared across all the searched frames are kept and the rest points which are considered as outliers are removed (Fig. 3.3 (a)-(b)). The assumption is that, for a camera with a sufficient frame rate, the non-occluded object parts should be visible within a certain number of frames.

According to the above mentioned point extraction approach, there are more points located on the target, i.e., in the central area of the detection box, and less points near the box boundaries. Since these retained points are usually with the same motion pattern as the target, the distance between them are nearly constant. Here it assumes that these points are generated by a Gaussian Mixture Model with K components. The number K here can be determined by the *Mean Shift* algorithm. In this algorithm, a circle window with a radius r_w (empirically set to one third of the smaller dimension of the target box) is initially placed on each feature point. For each window, its centroid is calculated with respect to the covered feature points. In the next step, each window is shifted to its centroid. Above steps are iterated until all window positions are fixed. In the end, duplicated windows are removed and the number K is exactly the number of remaining windows.

Parameters of the Gaussian Mixture Model can be estimated by the *Expectation Maximization (EM)* algorithm. Here we assume that the probability of a point $\boldsymbol{l}_i = [x_i, y_i]^\top$ drawn from the Gaussian Mixture Model is interpreted as

$$p(\boldsymbol{l}_i) = \sum_{k=1}^{K} \pi_k \mathcal{N}(\boldsymbol{l}_i | \boldsymbol{\mu}_k, \boldsymbol{\Sigma}_k), \tag{3.13}$$

where $\boldsymbol{\mu}_k$ and $\boldsymbol{\Sigma}_k$ respectively indicate the mean vector and covariance matrix of the k-th 2D Gaussian component. Term π_k denotes the corresponding weight. The EM can be divided into two steps: the Expectation and Maximization.

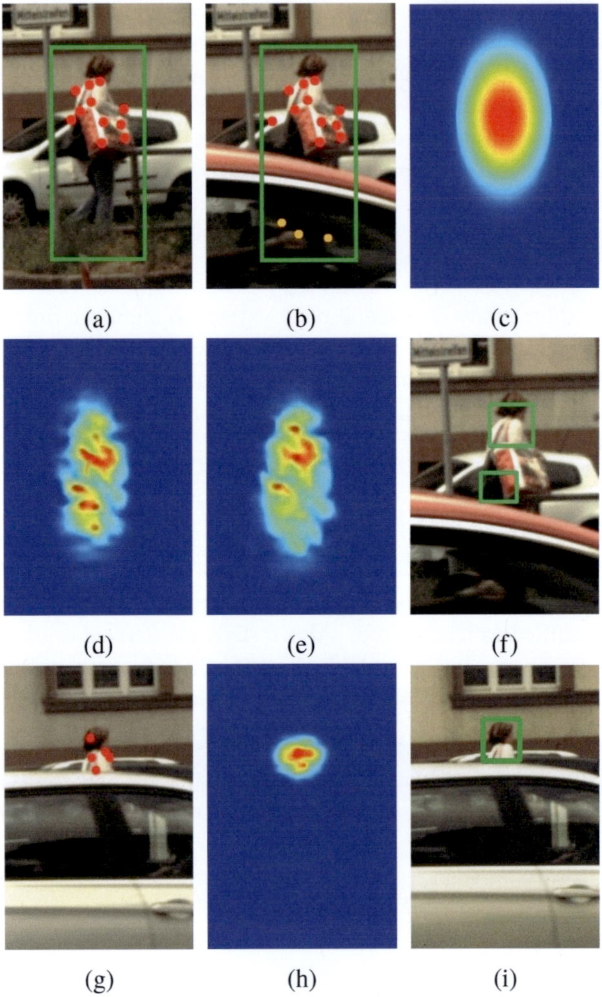

Figure 3.3: Image (a) and (b) show selected feature points in the green detection box with the color red and yellow respectively to denote inliers and identified outliers. Image (c) represents the initialized Gaussian filter, which is then multiplied with learned appearance model, represented in (d). The resulting energy map is shown in (e). Constructed part filters are displayed in (f). The last row shows an example with severe occlusion, in which only one part filter is utilized, as illustrated in (i). Image (g) and (h) respectively represent the inlier feature points and the resulting energy map.

In the first step, it calculates the probability of point $\boldsymbol{l}_i$ belonging to the k-th component, interpreted as

$$\gamma_{i,k} = \frac{\pi_k \mathcal{N}(\boldsymbol{l}_i | \boldsymbol{\mu}_k, \boldsymbol{\Sigma}_k)}{\sum_{k=1}^{K} \pi_k \mathcal{N}(\boldsymbol{l}_i | \boldsymbol{\mu}_k, \boldsymbol{\Sigma}_k)}. \tag{3.14}$$

Thus, the weight can be updated as

$$\pi_k = \frac{1}{N} \sum_{i=1}^{N} \gamma_{i,k}, \tag{3.15}$$

where N denotes the total number of points. In the next step, it estimates all component parameters by maximizing the summed log-likelihood

$$\max_{\boldsymbol{\mu}_k, \boldsymbol{\Sigma}_k} \sum_{i=1}^{N} \log \left(\sum_{k=1}^{K} \pi_k \mathcal{N}(\boldsymbol{l}_i | \boldsymbol{\mu}_k, \boldsymbol{\Sigma}_k) \right). \tag{3.16}$$

Both steps are iterated until the likelihood converges or the predefined iteration number is reached. For initialization of Gaussian components, we use the parameters estimated by the previously utilized Mean Shift algorithm (i.e., the mean location and covariance matrix of each point cluster). The weight π_k is initialized as equally distributed. In the experiment with severe occlusion, it is found that in most cases there is only $k = 1$ component and thus only the maximization step needs to be executed. The estimated parameters are further used to create Gaussian filters, which are multiplied with the learned object appearance model. In this way, the visible parts in the appearance model can be acquired. The appearance model only with non-occluded object area is then utilized as an energy map to define the number of part filters. An example of above procedure with $k = 1$ component is illustrated in Fig. 3.3 (c)-(e).

In the subsequent step, each part filter is initialized with a rectangular shape of a predefined size $W_p \times H_p$ and consecutively placed over the energy map. After the placement of each filter, the energy map is updated by removing its covered pixels. Each filter is put on the location where the sum of its covered pixel values (i.e., the energy) is maximized. This procedure is repeated until either the sum of remaining energy falls below a predefined threshold θ_E or the maximal filter number θ_F is reached. Therefore, the filter number is adapted to the current available appearance of tracked object, as shown in Fig. 3.3 (f).

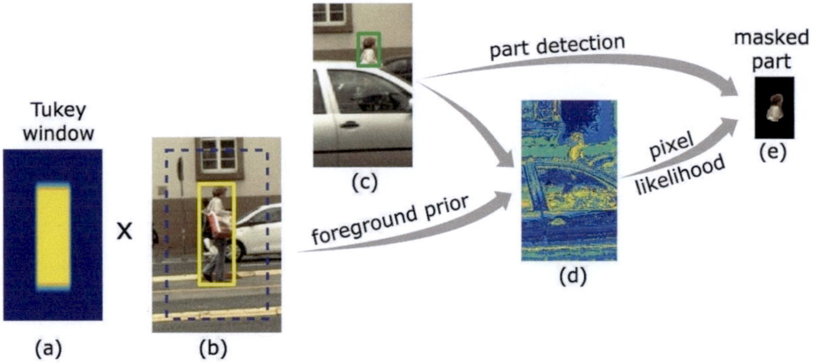

Figure 3.4: The masking process. From left to right, it respectively presents (a) the Tukey window, (b) the fore- and background region in the image, (c) the part detection, (d) the foreground likelihood illustrated by heatmap and (e) the yielded mask for visible parts.

3.2.4 Masking Process

Since occlusion is usually unpredictable, for an improperly selected size or shape of part filters, it is unavoidable to involve background objects during learning the appearance model of filters, which further prompts matching errors and even tracking failures. To alleviate this impact, a masking process is used for a more precise identification of visible target areas. In this approach, for each previously stored non-occluded image, both fore- and background region are defined for the target (respectively denoted by the yellow and blue bounding box in Fig 3.4 (b)). The foreground region is directly related to the current detection while the background region is twice as large as the foreground and is with the target situated at its center.

To extract the foreground region, a 2D Tukey-window with an equivalent size of the target is utilized, as illustrated in yellow in Fig 3.4 (a). The window is padded with zeros (denoted in blue) to the same size of the background region and then multiplied with the corresponding image patch. According to the definition of Tukey-window, its boundaries are smoothed by a cosine function to reduce the *leakage effect* of filtered signal in the frequency domain. Here the Tukey-window is used to suppress the selection of boundary pixels in the foreground region, because their probability belonging to the target is small. Thereafter, the background region is extracted by subtracting the foreground from the original image. For both regions color histograms are calculated and thenceforth normalized by the pixel number. These histograms are averaged over the saved images in the storage and then respectively packed into two vectors $\mathbf{h}_F$ and $\mathbf{h}_B$. Given each new detection of the object or its part (Fig. 3.4 (c)), the foreground likelihood for a pixel I_p is calculated by

$$L_F(I_p) = \frac{B_F(I_p)}{B_F(I_p) + B_B(I_p) + \epsilon}, \tag{3.17}$$

where term $B_F(I_p)$ and $B_B(I_p)$ respectively indicate the bin value in histogram $\mathbf{h}_F$ and $\mathbf{h}_B$ related to pixel I_p. Term ϵ is a small positive value to avoid division by zero. The likelihood of background pixel $L_B(l_p)$ can be calculated analogously. The calculated likelihood map L_F for the new image patch is illustrated in Fig. 3.4 (d) with bright colors to indicate high probabilities. Leveraging the part detection and the likelihood maps, it is possible to mask the visible area by a segmentation approach such as [89] (see Fig. 3.4 (e)).

The deployed approach [89] conducts segmentation in an EM-like fashion. Here a pixel is represented by a feature vector $\mathbf{v}_i$, which encodes the information of pixel color and coordinates. Each pixel is assumed to be drawn from a Gaussian Mixture Model with K' components. In this case we set $K' = 2$, because only two segments are required, which respectively represent the fore- and background. The probability of pixel descriptor $\mathbf{v}_i$ is interpreted as

$$p(\mathbf{v}_i) = \sum_{k=1}^{K'} \pi'_k \mathcal{N}(\mathbf{v}_i | \boldsymbol{\mu}'_k, \boldsymbol{\Sigma}'_k), \tag{3.18}$$

with $\boldsymbol{\mu}'_k$ and $\boldsymbol{\Sigma}'_k$ respectively denoting the mean vector and covariance matrix of the k-th component. The term π'_k represents the corresponding weight. In this segmentation approach, we only consider image patches representing the detected visible parts (denoted by the bounding box in Fig. 3.4 (c)). For initialization of both Gaussian components, fore- and background pixels are coarsely identified by comparison between variables L_F and L_B. Gaussian parameters are initialized based on these identified points. The weight π'_k is initialized as equally distributed. Similar EM-steps as described in Section 3.2.3 are then iterated to optimize the parameters of all components. Thereafter, pixels, which are with a greater probability belonging to the foreground than to the background, are considered as foreground pixels. In a further step, morphological operations (e.g., the closing and opening) are performed to fill small holes and to remove outlier points.

In the segmented image patch, all the background pixels are assigned with zero values. The filter size is then adjusted to the masked image area (with non-zero pixel values) by forcing the filter to cover at least 95% of the pixels of visible object parts. Within this covered image area (i.e., a bounding box), a conventional feature extraction approach such as the HOG is conducted to build appearance model for part filters.

3.2.5 Filter Management

Given a number of m part filters, by applying them on an input image $\mathbf{z}$, the same number of filter responses is acquired. As the evaluation of each part filter is independent from each other, they can be effectively parallelized, e.g., by a multi-threading process. For a fair evaluation on the discriminative power of various part filters, a weighting approach is adopted and interpreted as

$$\mathbf{f}^p(\mathbf{z}) = \sum_{l=1}^{m} c_l \mathbf{f}_l^p(\mathbf{z}(\boldsymbol{\Delta}_l)), \tag{3.19}$$

where $\mathbf{f}_l^p(\mathbf{z})$ is the evaluation response of part l. The shift vector $\boldsymbol{\Delta}_l$ aligns its response map to the object center. The weight c_l indicates the discriminative power of corresponding part filter, which is measured by aforementioned criteria, interpreted as

$$c_l = \frac{1}{c_\Sigma} \cdot \mathrm{PSR}_l \cdot \exp(-\mathrm{NOD}_l), \tag{3.20}$$

with the normalization factor

$$c_\Sigma = \sum_{l=1}^{m} \mathrm{PSR}_l \cdot \exp(-\mathrm{NOD}_l) \tag{3.21}$$

to guarantee that the weight sum equals 1. The target location is finally inferred by the peak point of the aggregated response map $\mathbf{f}^p(\mathbf{z})$. For each part filter, the classifier coefficient vector $\boldsymbol{\alpha}_l^t$ and the appearance model $\mathbf{x}_l^t$ at frame t are updated by the currently learned ones α_l and $\mathbf{x}_l$ in a similar way to Eq. (2.31), which is mathematically expressed as a linear interpolation

$$\begin{cases} \mathbf{x}_l^t = (1-\beta)\mathbf{x}_l^{t-1} + \beta\mathbf{x}_l & (3.22a) \\ \alpha_l^t = (1-\beta)\alpha_l^{t-1} + \beta\alpha_l & (3.22b) \end{cases},$$

where learning rate β is set to a small positive value. A similar interpolation approach with the same learning rate β is also utilized to update the fore- and background color priors $\mathbf{h}_F$ and $\mathbf{h}_B$.

Since the foreground color prior can coarsely identify the pixels belonging to the target in new frames (introduced in Section 3.2.4), an idea to further improve the inference precision of the full object tracker is to incorporate such information in the final response map $\mathbf{f}_\Sigma(\mathbf{z})$, which can be expressed as

$$\mathbf{f}_\Sigma(\mathbf{z}) = (1-\gamma)\mathbf{f}(\mathbf{z}) + \gamma\mathbf{f}_F(\mathbf{z}), \tag{3.23}$$

where $\mathbf{f}(\mathbf{z})$ indicates the original response map by evaluating the full object tracker on image sample $\mathbf{z}$ and parameter γ is a small positive merging factor. The response map $\mathbf{f}_F(\mathbf{z})$ is generated by matching the foreground color histogram of the whole target with that of each shifted version of sample $\mathbf{z}$. Both the full object tracker and its color priors are updated in the same fashion as Eq. (3.22). An example of merged responses is illustrated in Fig. 3.5.

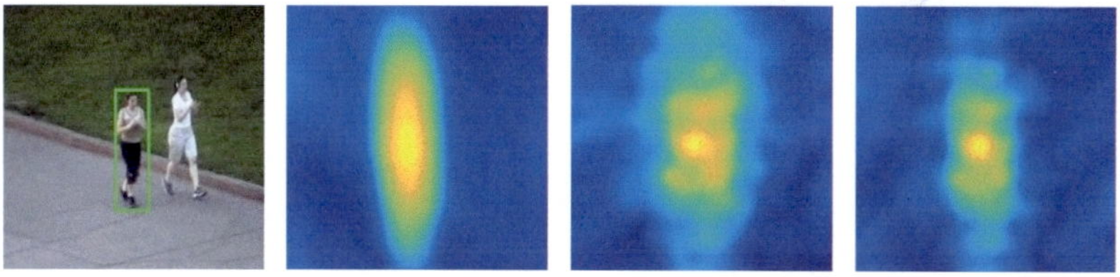

(a) (b) (c) (d)

Figure 3.5: Target inference in combination with the foreground color matching. Subfigure (a) shows the test image with the target denoted in green bounding box. Subfigure (b) to (d) respectively show response maps (displayed as heatmap) for foreground color matching, original full object tracker and the merged one according to Eq. (3.23). Image sample is from [167].

To alleviate the increased computational burden caused by additional part trackers, a dynamic filter management framework is adopted, shown in Fig. 3.6.

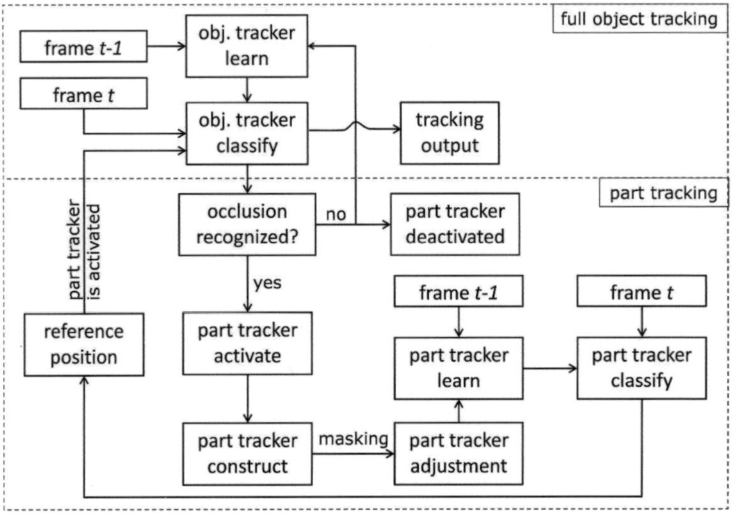

Figure 3.6: Illustration of the proposed tracker management framework. For each frame, the occurrence of occlusion is checked with respect to the criteria introduced in Section 3.2.2. In the case of occlusion, part filters are constructed and fine adjusted according to strategies in Section 3.2.3 and 3.2.4. During occlusion, only part trackers are updated while the full object tracker is kept unchanged. Simultaneously, the full object tracker is evaluated on the location inferred by part trackers. If the criteria for occlusion recognition are no longer satisfied, part trackers are shut down and the update of full object tracker is reactivated again.

In this framework, the full object tracker, which is learned from the previous frame, is applied on the current frame to identify the target. Simultaneously, the occlusion detection algorithm introduced in Section 3.2.2 is also executed for each frame. Once the occlusion is identified, part trackers are constructed according to strategies introduced in Section 3.2.3 and 3.2.4. The part filters are learned on previously saved frames and applied on the current frame to identify the visible parts. During occlusion, only part filters are updated while the full object tracker is kept unchanged. In the meanwhile, the full object tracker is evaluated on the location inferred by part filters. If the criteria introduced in Section 3.2.2 are no more satisfied in s consecutive frames, it assumes that the occlusion disappears. Thereupon, part filters are shut down and the full object tracker is activated again. In the meanwhile, the update of full object tracker is also reactivated. Since part filters are only constructed when the occlusion is detected and then deactivated for non-occluded cases, the tracking approach is efficient.

3.3 Evaluation

In this section, the proposed tracking approach is evaluated in comparison with several state-of-the-art trackers based on two datasets: the driving dataset [146] and the OTB benchmark [167]. Simultaneously the effectiveness of proposed occlusion recognition mechanism is verified and the superior performance of proposed tracking in dealing with other challenging scenarios is demonstrated.

3.3.1 Experimental Setup

To reveal the performance of occlusion identification process, two versions of the proposed approach are prepared, depending on whether the masking process is incorporated or not. Both versions are implemented by the basic KCF model [71] in C++ programs and respectively denoted as KCF_P and KCF_PM, where the letter “P” stands for integration of part filters and the letter “M” represents the utilization of the masking process. Additionally, we evaluate one more tracker called as KCF_PMP, with the last letter “P” to indicate the promoted inference by the deployment of foreground color matching for the full object tracker. For each version of the above employed trackers, their part filters share the same structure as the KCF. All of them use FHOG [50] and Color Names [40] as features. The performance of these proposed trackers is compared with the baseline model KCF as well as with seven other state-of-the-art trackers, i.e., Struck [65], DSST [36], MEEM [172], DPCF [1], RPT [99], MUSTer [73] and SRDCF [38]. Among them, the first three trackers are mainly based on a linear correlation filter. The DPCF and RPT employ explicit part modeling and are based on the KCF. The last two trackers are also CF-based but with implicit occlusion handling by enhanced learning strategies.

Parameters of all evaluated methods are consistent with their original papers. The test platform is a laptop of a quad-core CPU with a working frequency of

2.7 GHz and a memory of 8 GB. In the proposed approach, the thresholds are empirically set to $\theta_P = 7$ and $\theta_N = 0.04$. The frame number q and s are set to 4. The learning rate equals $\beta = 0.125$. The merging factor is set to $\gamma = 0.3$. The part filter is initialized in a square shape with its size equal to half of the smaller dimension of the object. The energy threshold equals $\theta_E = 0.5$ while the maximal filter number is $\theta_F = 2$. As the visible part is usually small due to severe occlusion, this number is found sufficient in the experiments.

The evaluation exactly follows the One-Pass Evaluation (OPE) protocol [167], where the tracker is triggered at the initial frame and both object location and size are estimated in subsequent frames. Tracking results are reported by the precision and the success plots. The first one is calculated according to the distance between centers of the target and its ground truth while the second criterion is based on their overlap ratio. Same as in [167], for a qualitative comparison, trackers are ranked at the threshold of a distance error of 20 pixels and an overlap ratio of 0.5.

3.3.2 Evaluation on Real Traffic Scenarios

Since this thesis is focused on object tracking applied in traffic scenarios, we firstly adopt the driving dataset [146] to evaluate the proposed approach. This dataset consists of video sequences captured by a vehicle with roof installed cameras driving in the city. The images are recorded with a frame rate of 10 frames per second (fps) and a resolution of 1200×712 pixels. The total recording process takes about 4 hours. These sequences are sorted into three groups. Each group consists of both labeled pedestrians and bicyclists. The minimal height of labeled object is 80 pixels. In the first group, more than 60% of an object can be observed in the image. Thus, it can be considered as a tracking baseline without or only with minor occlusion. The second group consists of sequences, in which pedestrians and bicyclists are partially occluded by street vehicles within a short period (up to 20 frames). The third one includes sequences in similar scenarios with a long time occlusion (up to 70 frames). In the last two groups, less than 40% of an object is observable, corresponding to the severe occlusion.

Verification of Criteria for Occlusion Recognition

In the proposed approach, two criteria are utilized, i.e., the PSR and NOD value, to recognize the abrupt appearance change, which is mainly due to severe occlusion. Since the criteria for occlusion recognition directly determine the activation of part filters and further influence the total tracking performance, the first experiment is aimed to verify its effectiveness. This experiment is conducted on the driving dataset with three comparison sets: two methods with each only integrated with one criterion, and the proposed method with both criteria employed. Only the related trackers, i.e., KCF_P, KCF_PM and KCF_PMP, are chosen for evaluation. Additionally, one more comparison set with the naive KCF is added as the baseline. Here only the results on the sequence groups with short term and long term severe occlusions are reported, because both criteria are rarely activated on sequences with minor occlusions, which will make the comparison less reasonable. The quantitative evaluation is based on the precision value at the distance error of 20 pixels and the success rate at the overlap ratio of 0.5, which follows the protocol of [167].

All the results are recorded in Table 3.1. It is clear that the naive KCF performs poorly in both sequence groups. Especially for long term severe occlusion, its precision value and success rate are under 40%. After integrated with part filters, even with only one criterion, its performance on both metrics are significantly improved. Generally, the KCF_PM outmatches KCF_P in terms of precision and success rate, which can be credited to the masking process, which provides a more precise identification of visible object areas. The KCF_PMP performs superior over KCF_PM, which implies the foreground color matching indeed benefits the target inference. In the short term occlusion group, the criterion NOD performs slightly better than PSR, which means the object appearance model is relative more sensitive to variations in this case. However, the highest score for both metrics is obtained by involving both criteria, which means the PSR criterion can still help NOD in recognizing the occurrence of occlusions. For the long time severe occlusion, scores of PSR and NOD are very close, which means significant changes on both the appearance model and filter response are captured. Still, the best performance is obtained when both criteria are incorporated, which demonstrates the effectiveness of combination of both metrics.

Occlusion	Method	Precision				Success rate			
		None	PSR	NOD	PSR+NOD	None	PSR	NOD	PSR+NOD
Short term severe	KCF_PMP	-	86.8	87.1	88.0	-	83.6	83.4	84.7
	KCF_PM	-	84.7	85.8	87.9	-	82.5	83.2	84.0
	KCF_P	-	84.2	84.9	86.5	-	81.3	82.4	83.2
	KCF	72.5	-	-	-	69.3	-	-	-
Long term severe	KCF_PMP	-	71.8	72.1	72.9	-	70.2	70.5	71.9
	KCF_PM	-	70.1	69.8	70.8	-	69.4	69.5	70.1
	KCF_P	-	68.4	68.5	69.2	-	68.0	67.7	68.7
	KCF	38.4	-	-	-	39.5	-	-	-

Table 3.1: Verification of criteria for occlusion recognition mechanism on the driving dataset with score values displayed in %.

Evaluation on Driving Dataset

In this experiment, the proposed approach is compared with the state-of-the-art trackers based on the driving dataset. Experimental results for each sequence group are presented in the form of precision and success plot illustrated in Fig. 3.7. It can be seen, in the case of minor occlusions, all tested methods exhibit a relative good performance (Fig. 3.7 (a)). It assumes that their appearance models are slightly influenced by the occlusion, thus the classification power remains stable. For the KCF_P tracker, its precision value is about 2% higher than the naive KCF approach, which proves the effectiveness of integration of part filters. By employing the masking process, a further gain of 1.6% is obtained by the KCF_PM tracker, which is credited to the pixel-level identification of visible object areas. Similar trend can also be seen in other plots. Generally, the KCF_PM and KCF_P outperform normal CF-based trackers like Struck, DSST and MEEM in most cases, which proves the effectiveness of employing part filters. Although RPT and DPCF also consist of part trackers, as the masking process is integrated in KCF_PM, the target identification by KCF_PM becomes more precise. This effect is more obvious with short term severe occlusions, where the KCF_PM tracker achieves the second highest precision (Fig. 3.7 (b)). Since in the first sequence the occlusion is little (Fig. 3.8 (1a)-(1c)), activation of part filters is very few. Thus, KCF_P and KCF_PM perform inferior to MUSTer and SRDCF (with enhanced classifier learning). However, by color matching, KCF_PMP outperforms all of them.

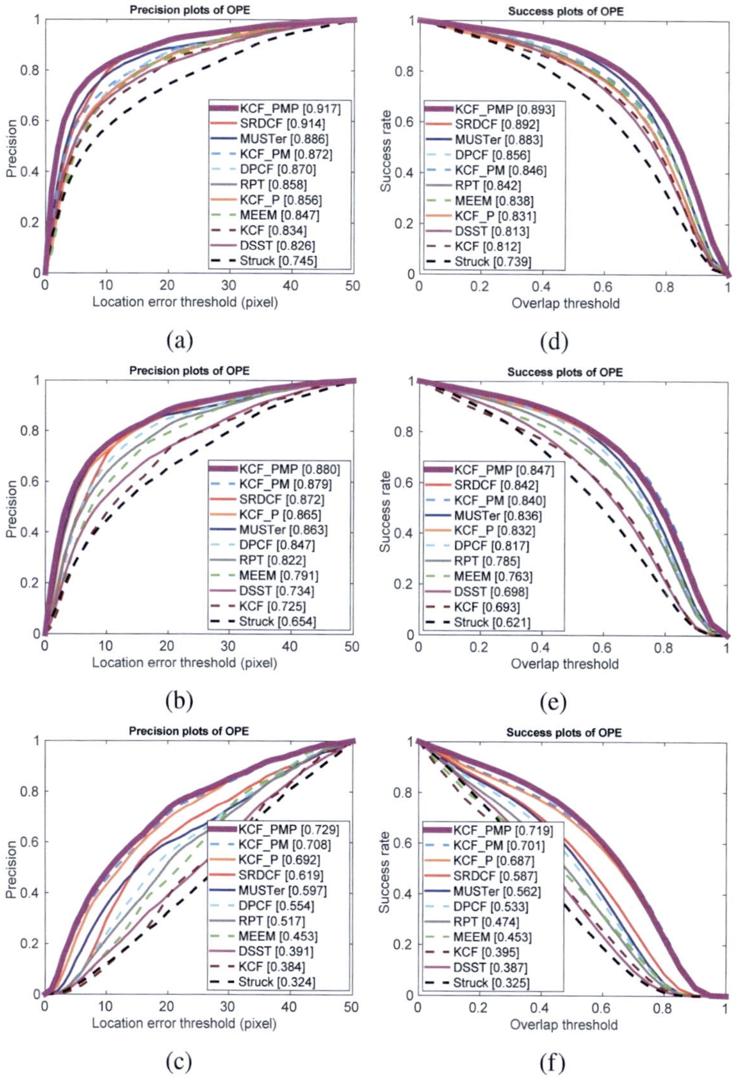

Figure 3.7: From top to bottom, the tracking results of the driving dataset on minor occlusions, short term severe occlusions and long term severe occlusions are presented respectively. For each case, both precision and success plots are presented. Plot of the proposed method is marked in bold. In the legend, the ranking list is displayed in a descending order.

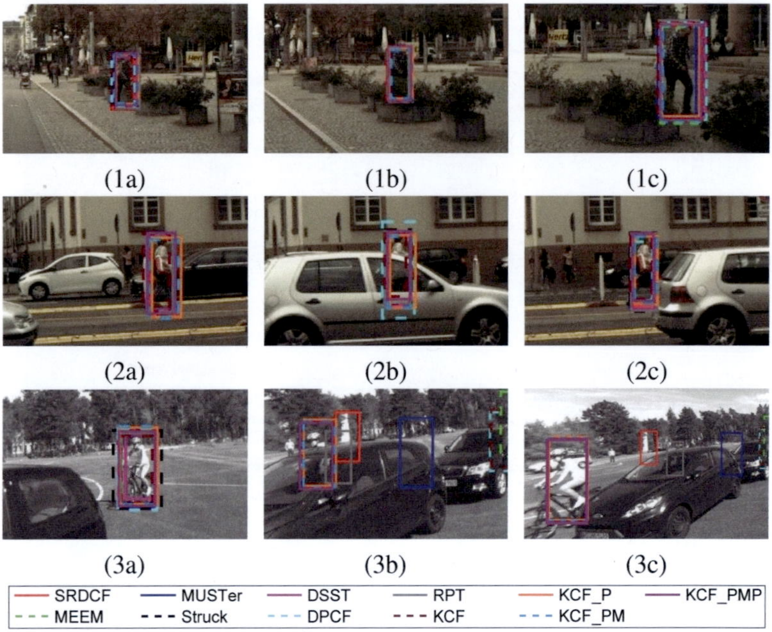

Figure 3.8: Three examples of the driving dataset with each displayed in one row. They respectively show the case of object tracking with minor occlusion, short time severe occlusion and long time severe occlusion.

As the occlusion becomes severe in last two sequence groups, part filters are activated more frequently, thus the accuracy of KCF_P and KCF_PM increases. In the case of long time severe occlusion, all trackers perform poorly except the ones integrated with part filters (Fig. 3.7 (e)). In these sequences, the maximal occlusion rate of an object is more than 85% and the average duration is about 70 frames. As the full object is for a long time unobservable, the appearance model of most of the trackers is contaminated by background objects and thus they lose the target right after the occlusion appears (Fig. 3.8 (3a)-(3c)). With exact awareness of observable parts, proposed approaches perform superior over them, especially over other kinds of part-based trackers. A similar trend can be seen in the corresponding success plot (Fig. 3.7 (f)).

3.3.3 Evaluation on General Tracking Tasks

Although the introduced solution is mainly proposed for object tracking in traffic scenarios, as the employed tracker model is not limited to specific object classes, its application in general tracking tasks should also be possible. To verify this assumption, we resort to another standard dataset, i.e., the OTB benchmark [167], including two subdatasets: OTB2013 and OTB100. The first one contains 50 sequences with observed objects of varied sizes, dynamics and heterogeneous classes including traffic participants as well as small objects like balls, books and toys. Both indoor and outdoor scenes are captured with challenges like occlusion and variation of motion or perspectives. The second dataset includes another 50 sequences with a similar setup as in the first one but involves more challenging scenarios. Both datasets are considered as difficult and widely tested in numerous works.

General Evaluation on OTB Benchmark

In the first experiment of this section, the general performance of all compared trackers (including the proposed KCF_P, KCF_PM and KCF_PMP tracker) on both OTB datasets is evaluated. Here the same evaluation protocol as in previous experiments is chosen and the tracking results of compared methods are reported in Fig. 3.9. It is obvious that for both datasets, by employing part filters, the precision of naive KCF approach is promoted by more than 7% in the KCF_P tracker (Fig. 3.9 (a)-(b)). And the success rate gain is more significant, about 9%. Although the performance of KCF_P is greatly boosted, it is lower than the top performed trackers, such as MEEM, MUSTer and SRDCF. Even integrated with the masking process in KCF_PM, the performance is still less comparable. The reason can be owed to the structure of KCF tracker itself, which is a straightforward implementation and strictly follows the learning rule of correlation filter. In contrast, MEEM chooses expert filter with less ambiguities from multiple snapshots. Although it is previously demonstrated not good at dealing with severe occlusion, it still shows robustness against fast motion, rotation, etc. As for the other two trackers, MUSTer switches tracking between short and long memories while SRDCF performs spatial regularization among training samples. These additional manipulations make them more flexible in handling various tracking conditions. Although the part filter is verified effective for handling occlusions, its power can be limited by simple model structure, e.g., in the KCF tracker, and cannot eliminate such great performance gap between those top performed ones. In comparison, by incorporating foreground color matching for the final response map, the tracker KCF_PMP takes the first place in the ranking for both datasets, which further proves that the foreground color is an important information and robust against variation of size, shape, orientation, etc.

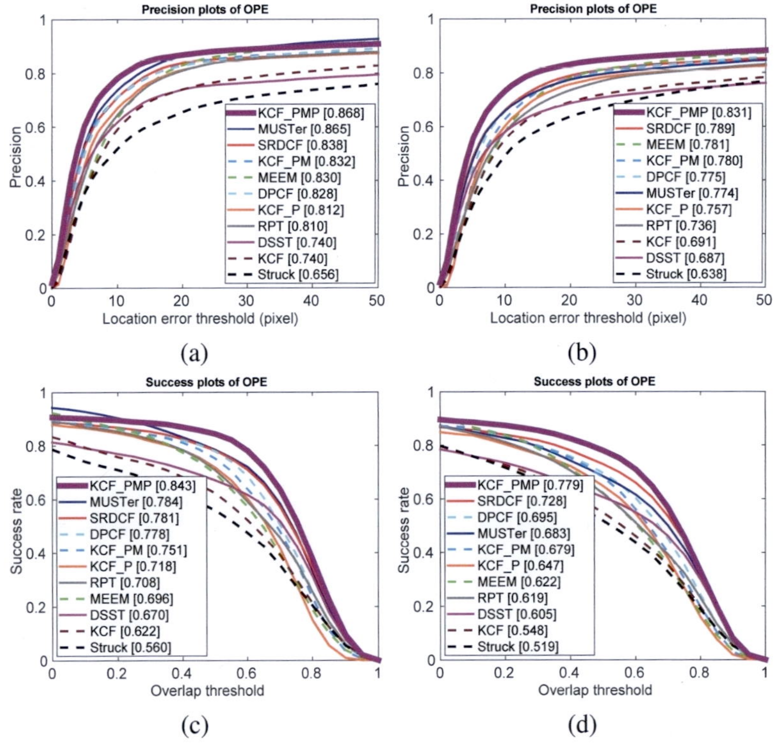

Figure 3.9: From left to right, tracking results of the OTB2013 and OTB100 dataset are presented respectively. Each column shows the corresponding precision and success plots. Plot of the proposed method is marked in bold. In the legend of each plot, the trackers are ranked in a descending order in terms of a location error of 20 pixels and an overlap ratio of 50%.

Attribute-based Evaluation

In a further experiment, based on provided sequence attributes in the OTB benchmark, the performance of the proposed approach is explored in more specific challenging scenarios with eight attributes covering occlusion, motion blur, illumination variation, in- and out-of-plane rotation, background clutter, scale variation and fast motion. For each tracker, both the precision and success rate in terms of a distance error of 20 pixels and an overlap of 0.5 with the groundtruth are respectively reported in Table 3.2 and Table 3.3.

	occlusion	motion blur	illumination variation	in-plane rotation	out-of-plane rotation	background clutter	scale variation	fast motion
SRDCF	84.3	78.9	76.1	76.6	81.8	80.3	77.8	74.1
MUSTer	85.3	69.5	79.5	79.9	85.0	83.1	81.7	69.5
DPCF	86.2	72.3	80.4	75.4	80.7	81.1	75.3	70.1
RPT	76.3	78.6	81.0	80.9	79.6	84.2	78.9	74.7
MEEM	79.9	71.5	76.6	80.0	84.0	79.7	78.5	74.2
DSST	70.6	54.4	73.0	76.8	73.6	69.4	73.8	51.3
Struck	56.4	55.1	55.8	61.7	59.7	58.4	63.9	60.3
KCF	74.9	65.0	72.9	72.5	73.0	75.2	67.9	60.2
KCF_P	78.2	70.9	74.7	76.4	79.9	78.1	73.3	68.8
KCF_PM	83.6	73.3	76.2	77.7	81.8	79.5	78.3	71.8
KCF_PMP	**86.6**	**80.4**	**81.5**	**84.8**	**86.1**	**84.5**	**81.9**	**81.3**

Table 3.2: Attribute-based evaluation. Each column represents the precision values (displayed in %) of compared methods for one attribute. The name of proposed approach and the best value are displayed in bold.

According to evaluation results, it is clear that the employment of part filters and masking process can improve the performance of KCF tracker in all tested cases with a gain of 2 ~ 10%. Particularly for the occlusion attribute, the KCF_PM tracker achieves a high precision value (Table 3.2), which is comparable with other well performed trackers (e.g., SRDCF, MUSTer), further proving the advantage of part modeling approach. Still, the best performance is achieved by the KCF_PMP model which is integrated with the color matching.

	occlusion	motion blur	illumination variation	in-plane rotation	out-of-plane rotation	background clutter	scale variation	fast motion
SRDCF	79.0	76.2	70.1	70.9	74.0	71.5	71.2	71.1
MUSTer	76.3	66.8	73.6	69.1	74.2	75.0	70.4	65.1
DPCF	79.9	66.6	74.7	69.7	74.6	75.6	68.2	63.8
RPT	64.4	72.0	68.3	70.9	68.2	75.2	63.0	69.3
MEEM	67.8	66.0	63.8	65.0	67.9	73.7	57.0	68.1
DSST	64.6	52.8	68.1	67.9	64.2	62.7	64.0	50.3
Struck	49.3	51.9	49.2	52.9	50.7	54.5	47.1	56.7
KCF	61.7	59.6	58.2	61.5	60.7	67.3	47.7	55.7
KCF_P	69.8	65.6	65.2	66.7	67.0	69.6	62.2	63.7
KCF_PM	76.4	68.5	68.2	69.3	70.9	71.4	68.2	67.1
KCF_PMP	**82.7**	**76.9**	**77.8**	**81.0**	**82.1**	**79.6**	**77.5**	**78.3**

Table 3.3: Attribute-based evaluation. Each column represents the success rates (displayed in %) of compared methods for one attribute. The name of proposed approach and the best value are displayed in bold.

Although part filters and masking process are originally designed for handling occlusion, it can be noticed that they can also benefit other scenarios such as unexpected motion, rotation, scale and illumination variation, etc. In such cases, the object appearance undergoes significant changes, which are equivalent to the effect by occlusion and thus trigger the activation of part filter. The part filters are normally placed on the most discriminative object areas while the masking process further boost the localization precision by removing background pixels. Hence, in those scenarios, the appearance model becomes more accurate and therefore the classifier learns better. This demonstrates the outstanding generalization ability of the proposed approach in heterogeneous tracking tasks, which is qualitatively illustrated with some examples in Fig. 3.10 and 3.11.

3.3.4 Runtime Performance Analysis

Here the average runtime performance of tested approaches on both datasets is reported in Table 3.4. The fastest approach is the naive KCF with a speed of slightly less than 40 fps. It is followed by DSST with a big gap of about 10 fps. On the third place is the KCF_P with a speed of 22.7 fps, due to the fact that the processing is greatly slowed down by training additional part filters. In comparison, the overhead of masking process is little, thus the KCF_PM is only about 2 fps slower. The foreground matching is also fast, only taking about less than 10 milliseconds (ms). Thus, the speed of KCM_PMP is 17.5 fps. Nevertheless, such speed is still acceptable in most real-time applications and is much faster than other well-performed trackers such as MUSTer and SRDCF. Moreover, the corresponding precision gain makes the KCF_PMP tracker outperform all state-of-the-art approaches in the experiments. However, there is still chances for further speeding up the proposed approach, e.g., by utilizing multi-threading for part filters or implementation on high performance hardwares such as high-end GPUs.

Method	Struck	DSST	KCF	MEEM	MUSTer	SRDCF	DPCF	RPT	**KCF_P**	**KCF_PM**	**KCF_PMP**
speed (fps)	2.3	27.0	37.1	5.3	0.8	1.7	5.47	3.62	22.7	20.4	17.5

Table 3.4: Evaluation of tested approaches on the runtime performance.

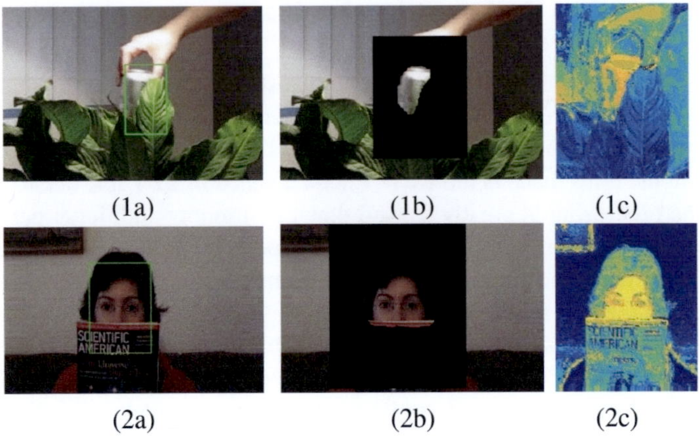

Figure 3.10: Two examples of occluded object from the OTB dataset with each displayed in one row. For each example, from left to right it respectively shows the recognition of full object (marked in a green bounding box), pixel-level part identification and foreground color prior (displayed in a heatmap and corresponding to the black rectangular area in the middle image).

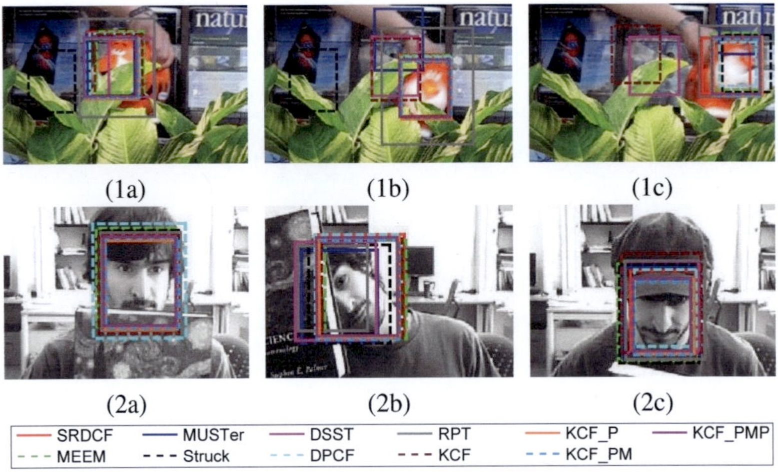

Figure 3.11: Tracking results of two sequences from the OTB benchmark with attributes of occlusion, fast motion, rotation, scale variation, etc.

4 Tracking with Deteriorated Vision

Visual tracking approaches represented by the correlation filter and its descendants have shown outstanding performance in general tracking tasks. Leveraging well-elaborated strategies, they can even handle complex cases such as severe occlusion, as demonstrated in Chapter 3. Despite such progress, their applications, discussed up to now, are mostly related to scenarios with relative good vision conditions. The adversarial problem - tracking with deteriorated vision, which is rarely considered in these works, however, is a challenging problem persistently confronted by the research community. Regarding the traffic scenario, which is the focus of this thesis, the referred problem frequently occurs in cases where the vision is significantly weakened by the low illumination (e.g., at night or in the fog) or by adverse weather factors (e.g., raindrops). This can yield great challenges for vision-based ADAS or automated driving systems, where instances, especially vehicles, in the surrounding of the ego-car may not be accurately identified due to the deteriorated vision, which can further lead to false maneuver decision of the ego-vehicle and thus result in traffic accidents.

For a more comprehensive exploration on this topic, the problem of vision deterioration mentioned here is coarsely categorized into two subgroups: constant and temporally varying vision deterioration. The first one is generally caused by poor illumination conditions like during the night or in the fog. In such scenarios the main problem for tracking tasks is the degraded contrast between the background and foreground objects, as reported in [26]. In low exposed images, the ordinary visual features utilized for object recognition, e.g., the size, shape and color information of an object, suffer from various fading effects and are only partially available [144]. Especially for a vehicle, its most distinguishable parts are thus the brightest ones, which are usually from the head- and taillights, turn signals, warning lights and stripes, or other reflection areas (Fig. 4.1 (a)-(b)). The second subproblem is mostly related to adverse weather factors, e.g., the rain. In such case the camera imaging can be interfered by unclear vision condition of the windscreen[1] such as mist or water drops. Although the windscreen can be cleared by devices like rain wipers, it still leads to short-time visual contamination of tracked targets by scattered bright areas in the image (Fig. 4.1 (c)-(d)). The corrupted object appearance can be learned by the tracker, which further leads to tracking drifts and even tracking failures.

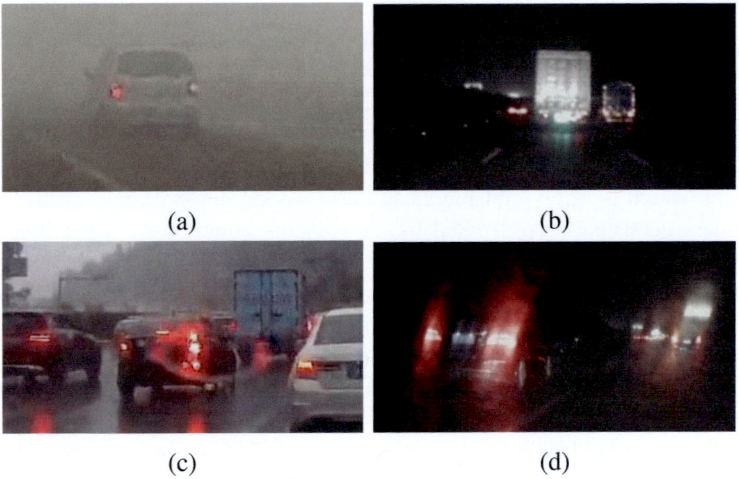

Figure 4.1: Image samples of deteriorated vision from [25, 144]. Image (a) represents a silver car with faded contrast against the background in the fog. The only recognizable parts are its taillights and rear window. Image (b) shows two trucks which are distinguishable by the white trunk and the warning stripes respectively. Images (c)-(d) represent the visual contamination in the image by raindrops on the frontal windscreen, with the camera mounted behind it. The light from vehicle lamps disperses to several bright areas in the image, which are a lot bigger than their original size. This dispersed light contaminates other vehicle parts, making the recognition of tracked object much more difficult.

The first subproblem requires the tracker to seek the most discriminative visual feature of an object while the second one demands outlier rejection for training samples. With both points in mind, in this chapter, a novel tracking solution based on the KCF tracker is introduced yet with an improved learning approach

[1] Here we assume that cameras are installed directly behind the windscreen which is a common setup for most of current ADAS and automated driving systems.

to directly tackle these problems in a visual feature level, which is published in the work [144]. The organization of this chapter is organized as follows: Firstly, we introduce state-of-the-art tracking researches on dealing with deteriorated vision condition. Afterwards, we present the procedure in dealing with constant vision deterioration by assembling different feature channels into several kernelized experts and utilize their estimated reliabilities to build appearance models. Thereafter, a temporal optimization procedure is introduced to remove outliers and keep the most reliable training samples. In following experiments, we show that the classifier, trained by both procedures, not only focuses on the discriminative image features, e.g., in low illuminated cases, but also is free from visual contamination by unreliable object candidates, e.g., caused by adverse weather factors. As the major computation task can be transferred into frequency domain, we also demonstrate that the proposed tracker can provide real time performance.

Since the proposed approach is mainly about learning robust classifiers for a target against the deteriorated vision condition, we only focus on the discussion about single object tracking in this chapter. Here we omit its application in cases of multi-object identification, because such topic is more related to association algorithms, which are addressed in the next chapter.

4.1 State of the Art

In this section we introduce state-of-the-art tracking works that deal with constant and temporal varying vision deterioration problems. Since solutions related to the first subproblem are much more diversified, we further sort its related works into two branches, i.e., enhancing image qualities and improving object recognition algorithms.

Enhancing Image Quality under Low Illumination

As previously discussed, the constant vision deterioration in traffic scenarios is usually caused by the low illumination condition, resulting in degraded visual contrast between foreground and background objects. Thus, an intuitive idea to solve such problem is to improve the quality of captured images. Typically,

O'Malley et al. in [118] employ a high dynamic range (HDR) camera with a specific hardware configuration to control the exposure and color processing functions at night. By adapted white balance, they segment rear lamps of a vehicle in the HSV color space and track them with Kalman filters. Although their approach shows robustness against changes in road environments, difficulty still occurs in estimating the size of a vehicle only based on the information of tracked rear lamps. Kim et al. in [86] present another approach by adopting a sonar senor in addition to a normal camera. They interpret the received ultrasonic signals as an additional chromatic channel to improve the contrast in low-exposed images. However, their ultrasonic sensor only has a very coarse resolution and can only measure instances within a limited range, i.e., about 10 meters. In the recent research [64], an advanced camera system, which fires ultrashort laser impulses for each pixel, is tested in a foggy scene. Their penetration depth is demonstrated to be superior over human vision yet especially for objects with small distances.

Aside from specific hardware configurations, post-processing of low-light images is also preferred in plenty of research works. Typical procedures to enhance the contrast of image is the histogram equalization and gamma correction. The first one balances the distribution of pixel value within the whole image while the other one shifts dark pixels to higher brightness. Such procedures are usually followed by denoising operations to suppress the noise amplified in dark regions. For instance, Malm et al. in [107] deploy an adaptive filtering with a kernel which is wide for homogeneous regions and narrow on pixels of sharp edges. Thus, they are able to efficiently reduce image noise while preserve object contours. In [69], Hasinoff et al. replace non-confident patches of one reference frame by confident ones extracted from alternate multiple frames. By such alignment and merging process, they achieve a high-quality photograph for a specific scene from an under-exposed image sequence. Deep learning approaches are also utilized by Chen et al. in [24] to learn the contrast enhancement and denoising in an end-to-end fashion, yielding boosted performance. However, its computation load is still high, which is also a common case for other above mentioned methods.

Improving Object Recognition under Low Illumination

To the best of our knowledge, there are really few works about systematical investigation of object tracking under low illumination condition. Most related works are about tracking a specific object class, especially the vehicle, during nighttime. As visual features are strongly weakened in such case, most of the research works prefer to search bright areas of the target in the image. Representatively, Chen et al. in [26] take advantage of multilevel histogram thresholding to segment bright image regions resulted by head- or taillights of nearby vehicles during nighttime. The presence of a vehicle can be easily verified by a set of predefined rules based on these segments, which is demonstrated robust against various low-illumination conditions. A similar approach is employed by Robert et al. in [127] yet decomposed into two stages. The first stage is to detect vehicles by identifying their lamps through searching bright blobs while the second stage verifies these hypotheses via a decision tree classifier testing on other appearance features such as windscreens. This two-stage detection framework is followed by a Kalman filter module and shows stability for vehicle tracking in low-exposed images. Aggregated classifiers like AdaBoost are also exploited by Zou et al. in [176] for nighttime vehicle lamp detection. Additionally, they use motion information to help search lamp pairs. Since vehicle lights usually occur in pairs, this pairing process significantly improves the recognition accuracy.

The common shortcoming of above approaches is that although vehicle lamps appear as quite discernible in the night, they may not cover all the observable areas of a vehicle. Other parts with clear form or color are also worth being considered as appearance features to improve the tracking performance. However, such sort of deep digging on available visual information of a tracked target is rare to be found in current research works. Aside from above mentioned approaches, object recognition at night is also guided by additional sensors in other researches. Representative examples are the works [53, 92], in which thermal senors are adopted to provide prior knowledge for the presence of an object and thus to narrow down the search region in the image. Although they show a more efficient and precise classification result due to shrinked region of interest (ROI), the performance of their thermal sensors can still be interfered by unexpected heat sources such as bonfires.

Suppressing Corrupted Training Samples

The corruption of image samples is a common problem in the visual tracking domain and mostly caused by the abrupt change of observed scenes such as the intrusion of unexpected objects, significant variation of environmental illumination and the influence of adverse weathers. Since corrupted samples alway provide an inaccurate or even a fake representation of the target, involving them into the training set, can misguide the classifier on estimating the distribution of positive samples in the feature space, and further lead to increased classification errors. Hence, the core idea to tackle such problem is to distinguish between corrupted samples and their true positives. This point is already addressed in some recent tracking works. For instance, Nebehay et al. in [115] break down the target into tiny parts which are represented by key points. They match these points frame by frame via optical flow and verify their correspondences through the geometry compatibility between points including their relative distances and orientations. Although image samples disobeying such consensus can be easily discovered, key point searching can be difficult especially in textureless images. In the work [141], Supancic et al. iteratively revisit previous frames and choose the "good" ones as training samples which minimize the loss of learning function. Despite a robust appearance model learned by their tracker, it is extremely time-consuming due to the repeated evaluation on previous frames.

In other works, approaches joining both the sample outlier rejection and tracker learning are more preferred by researchers. Representatively, Li et al. in [97] employs a CNN model to classify the target and introduce "noised labels"[2] to model the reliability of samples based on the quality of detection responses. Despite the robust temporal sampling achieved, their computational cost is relative high. Unlike that, correlation filters is deployed by Danelljan et al. in [37]. They reformulate the learning procedure to jointly optimize both the target appearance model and the sample weights. Albeit the corrupted samples are greatly down-weighted, as the circular data structure is damaged in their approach, it is still inappropriate for real-time applications.

[2] In [97], due to prediction error, false positives or unreliable samples are inevitably incorporated into the training data as positive samples. Such contamination of training data is considered as "label noise". In [97], the reliability of each sample label is considered in the training process by augmenting the loss function with additional terms.

In contrast to those above approaches, the tracker presented in this chapter is mainly based on the KCF model and maintains the circular data structure. Hence, the fast calculation in frequency domain is enabled. Furthermore, it incorporates a joint optimization over feature channels and temporal training samples. Thus, it can handle both low illumination conditions and corrupted sample images.

4.2 Tracking with Joint Reliability Estimation

In this section, a novel tracking approach is introduced to deal with deteriorated vision condition. This proposed framework is built on the KCF tracker (with a Gaussian kernel) due to its high tracking accuracy and fast processing speed. To handle the low illumination condition, in this approach, appearance features of the target are decomposed into different kernelized experts. Based on the reliability of each expert, the best features are employed to build appearance models. By further estimating the reliability of training samples in the time domain, the classifier is forced to focus on the most confidential samples and thus down-weight the corrupted ones.

For a simplified analysis yet without loss of generality, we take tracking at night as the main example to investigate the approach in dealing with low illumination condition in the following part. However, through experiments, we still demonstrate the introduced approach is able to handle other low illuminated cases such as the foggy ones, which are presented later.

4.2.1 Channel-wise Reliability Estimation

From previous chapters, we know that the KCF tracker employs the kernel function to map sample features into a higher dimensional space to generate a more distinguishable distribution and thus to facilitate the learning of the classifier. However, we have only explored the case for single-dimensional features and implied that it can be generalized to multiple channel features. For a better understanding of the merits and shortcomings of KCF, we explicitly derive its multi-channel form right here.

Since the image sample $\mathbf{x}$ is considered to be consisting of multi-channel features, it can be redefined as $\mathbf{x} = [\mathbf{x}^1; \ldots; \mathbf{x}^C]$ with a total channel number C. Leveraging the multi-channel evaluation form of the correlation filter (2.25) and the Gaussian kernel definition (3.2), the kernelized dot-product of two samples $\mathbf{x}_i$ and $\mathbf{x}_j$ can be reinterpreted as

$$\kappa(\mathbf{x}_i, \mathbf{x}_j) = \exp\left(-\frac{1}{\sigma^2}\sum_{c=1}^{C} h(\mathbf{x}_i^c, \mathbf{x}_j^c)\right) \tag{4.1}$$

with respect to the channel-wise operation

$$h(\mathbf{x}_i^c, \mathbf{x}_j^c) = \|\mathbf{x}_i^c\|^2 + \|\mathbf{x}_j^c\|^2 - 2\mathbf{x}_i^{c\top}\mathbf{x}_j^c. \tag{4.2}$$

Leveraging this formulation, the multi-channel form of kernelized convolution between sample $\mathbf{x}$ and $\mathbf{z}$ can be rewritten as

$$\mathbf{k}^{\mathbf{xz}} = \exp\left(-\frac{1}{\sigma^2}\sum_{c=1}^{C} \mathbf{h}(\mathbf{x}^c, \mathbf{z}^c)\right) \tag{4.3}$$

with

$$\mathbf{h}(\mathbf{x}^c, \mathbf{z}^c) = \|\mathbf{x}^c\|^2 + \|\mathbf{z}^c\|^2 - 2\mathbf{x}^c * \mathbf{z}^c. \tag{4.4}$$

According to above equations, it can be seen that the dot-product (or convolution) for each channel is conducted separately and thereafter accumulated in the final response. Although such a form brings the possibility for fast calculation, e.g., by employing parallelization approaches, it implicitly assumes that all the feature channels are regarded with equal contributions in measuring the similarity between image samples $\mathbf{x}$ and $\mathbf{z}$. Although this assumption works well for object tracking in scenarios of good lighting condition, it can be troublesome to handle object tracking in low illuminated cases, e.g., during nighttime. As each feature channel may encode a unique type of visual information, they can suffer from various fading effects caused by low illumination condition (Fig. 4.1 (b)). Thus, equally treating them can weaken the classification power of discriminative channels especially by averaging their responses. The matching results can then become vulnerable to the noise from non-discriminative channels. Moreover, features from various channels may be represented in different scales. A direct response accumulation can also induce unbalanced weighting of different feature types.

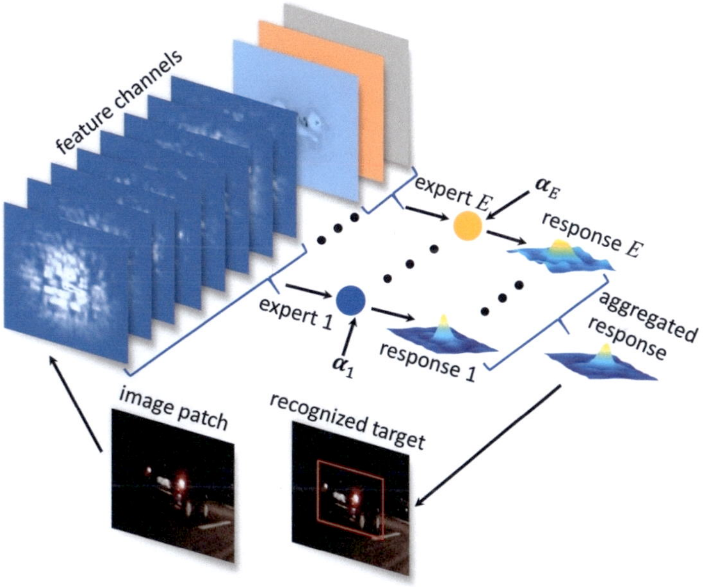

Figure 4.2: The classifier is decomposed into a number E of kernelized experts with each focusing on a limited number of feature channels. The filter response of each expert is weighted by its corresponding coefficient vector and aggregated into the final response map, with its peak to indicate the inferred location of the target.

To tackle these problems, an intuitive idea is to weight different feature channels or at least their convolution responses. With this inspiration, the evaluation result $\mathbf{f}(\mathbf{z})$ can be decomposed into a set of sub-evaluation results $\mathbf{f}_e(\mathbf{z})$ and thus formulated as

$$\mathbf{f}(\mathbf{z}) = \sum_{e=1}^{E} \mathbf{f}_e(\mathbf{z}) \tag{4.5}$$

with a total function number E and each sub-evaluation is in a form of

$$\mathbf{f}_e(\mathbf{z}) = \mathcal{F}^{-1}(\hat{\mathbf{k}}_e^{\mathbf{xz}} \odot \hat{\alpha}_e), \tag{4.6}$$

where corresponding parameters of evaluation function $\mathbf{f}_e(\cdot)$ are denoted by subscript $_e$. The vector $\hat{\alpha}_e$ here is more like a weighing factor to weight the convolution response $\hat{\mathbf{k}}_e^{\mathbf{xz}}$. In this concept, each sub-function is forced to only focus on a small number C_e of feature channels (Fig. 4.2), which encode visual information of the same type or represented in the same scale. Thus, the corresponding kernelized convolution $\mathbf{k}_e^{\mathbf{xz}}$ can be restricted to those features, expressed as

$$\mathbf{k}_e^{\mathbf{xz}} = \exp\left(-\frac{1}{\sigma^2}\sum_{c=1}^{C_e}\mathbf{h}(\mathbf{x}^c, \mathbf{z}^c)\right) \quad s.t. \quad \sum_{e=1}^{E} C_e = C. \tag{4.7}$$

Since each sub-function $\mathbf{f}_e(\cdot)$ only considers a very limited number of feature channels and shares a similar structure with the naive KCF tracker, $\mathbf{f}_e(\cdot)$ is called as kernelized expert. Recall that in Eq. (4.6) the convolution result $\mathbf{k}_e^{\mathbf{xz}}$ is further weighted by the coefficient vector $\boldsymbol{\alpha}_e$ in Fourier domain, thus the resulted filter response can be considered as a kind of reliability measurement, with great sharp values assigned to discriminative experts while non-discriminative ones represented by small flatten values (Fig. 4.3). In this way, feature channels can be fairly aggregated in terms of their classification power. Therefore, the rest problem is to learn a feasible coefficient vector $\boldsymbol{\alpha}_e$ for each expert $\mathbf{f}_e(\cdot)$.

In light of the Fourier formulation of correlation filter (2.21) and the kernelized evaluation function (2.30), the learning of naive KCF can be interpreted in frequency domain, which is

$$\arg\min_{\hat{\alpha}} ||\mathrm{diag}(\hat{\mathbf{k}}^{\mathbf{xx}})\,\hat{\alpha} - \hat{\mathbf{y}}||^2 + \lambda\,\hat{\alpha}^H \mathrm{diag}(\hat{\mathbf{k}}^{\mathbf{xx}})\hat{\alpha}. \tag{4.8}$$

However, the above form only considers sample $\mathbf{x}$ evaluated by a single classifier. For evaluation of the case of multiple experts, problem (4.8) should be rephrased as

$$\begin{aligned}&\arg\min_{\hat{\alpha}_e} ||\sum_{e=1}^{E} \mathrm{diag}(\hat{\mathbf{k}}_e^{\mathbf{xx}})\,\hat{\alpha}_e - \hat{\mathbf{y}}||^2 + \lambda \sum_{e=1}^{E} \hat{\alpha}_e^H \mathrm{diag}(\hat{\mathbf{k}}_e^{\mathbf{xx}})\hat{\alpha}_e \\ &= \arg\min_{\hat{\mathbf{a}}} ||\hat{\mathbf{K}}_1\hat{\mathbf{a}} - \hat{\mathbf{y}}||^2 + \lambda\hat{\mathbf{a}}^H\hat{\mathbf{K}}_2\hat{\mathbf{a}}\end{aligned} \tag{4.9}$$

with reshaped vector $\hat{\mathbf{a}} = [\hat{\alpha}_1; \ldots; \hat{\alpha}_E]$, matrices $\hat{\mathbf{K}}_1 = [\text{diag}(\hat{\mathbf{k}}_1^{\mathbf{xx}}), \ldots, \text{diag}(\hat{\mathbf{k}}_E^{\mathbf{xx}})]$ and $\hat{\mathbf{K}}_2 = \text{diag}([\hat{\mathbf{k}}_1^{\mathbf{xx}}; \ldots; \hat{\mathbf{k}}_E^{\mathbf{xx}}])$. Here the equal sign indicates the equivalence between those two optimization problems.

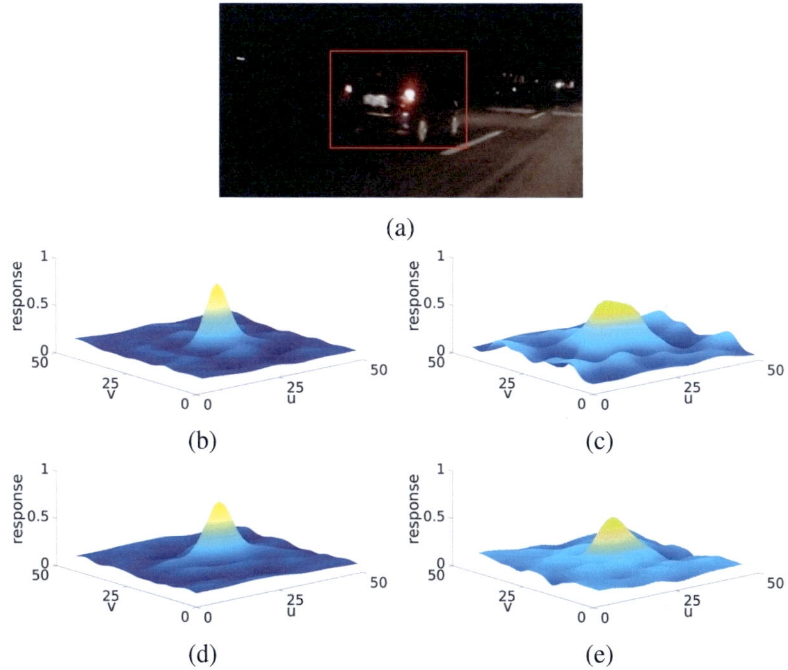

Figure 4.3: In this example, the appearance model comprises of two feature types: the FHOG features [49] and the color attributes [40], which are respectively integrated into expert 1 and 2. The tracked target is denoted by a red box in image (a). Image (b)-(c) show the filter responses for both experts. As the color of the vehicle undergoes great fading effect by low illumination, the gradient features turn out be more discriminative, which is implied by the sharp peak in the response map of expert 1. Accordingly, the response aggregated from both experts in image (d) also exhibits a much sharper shape than that of the ordinary KCF response in image (e).

Since the kernel function is equivalent to a mapping function, the vector $\mathbf{k}_e^{\mathbf{xx}}$ can be considered as the autocorrelation for specific feature channels of sample $\mathbf{x}$, which are interpreted in a higher dimensional space (cf. Eq. (2.27)). According to the Wiener-Khinchin Theorem [23], its Fourier transform $\hat{\mathbf{k}}_e^{\mathbf{xx}}$, only contains non-negative real values. Such conclusion is also valid for matrices $\hat{\mathbf{K}}_1$ and

$\hat{\mathbf{K}}_2$, as their non-zero elements only come from the vector $\hat{\mathbf{k}}_e^{\mathbf{xx}}$. By calculating second-order derivatives of objective function in (4.9), its Hessian matrix can be obtained as

$$\mathbf{H} = \hat{\mathbf{K}}_1^H \hat{\mathbf{K}}_1 + \lambda \hat{\mathbf{K}}_2. \tag{4.10}$$

In this equation, the first term is a Gram matrix due to that it only consists of non-negative real values. Thus, it is positive semi-definite. Since the second term $\lambda \hat{\mathbf{K}}_2$ is a diagonal matrix consisting of non-negative values, the Hessian matrix $\mathbf{H}$ is also positive semi-definite. In this way, problem (4.9) is semi-convex with respect to vector $\hat{\mathbf{a}}$. However, by well-designed features, the case that elements of vector $\mathbf{k}_e^{\mathbf{xx}}$ all equal zero can be avoided. Hencc, the Hessian matrix $\mathbf{H}$ becomes positive definite and the problem (4.9) is convex. Therefore, the optimum can be obtained at the point with zero derivatives, which can be formulated as $\mathbf{H}\hat{\mathbf{a}} - \mathbf{b} = \mathbf{0}$ with

$$\begin{cases} \mathbf{H} = \hat{\mathbf{K}}_1^H \hat{\mathbf{K}}_1 + \lambda \hat{\mathbf{K}}_2 & \text{(4.11a)} \\ \mathbf{b} = \hat{\mathbf{K}}_1^H \hat{\mathbf{y}} & \text{(4.11b)} \end{cases}.$$

For a better formulation, this equation set can be rearranged as

$$\mathbf{H}\hat{\mathbf{a}} = \mathbf{b}. \tag{4.12}$$

To solve this expression of vector $\hat{\mathbf{a}}$, the approach of *Successive Over Relaxation (SOR)* is exploited here. In this method, matrix $\mathbf{H}$ is decomposed in the form of $\mathbf{H} = \mathbf{D} + \mathbf{L} + \mathbf{U}$ with a diagonal matrix $\mathbf{D}$ as well as the strictly lower and upper triangular matrix $\mathbf{L}$ and $\mathbf{U}$. The approximation error of solution $\hat{\mathbf{a}}$ can be iteratively reduced by proceeding following operation

$$(\mathbf{D} + \omega\mathbf{L})\hat{\mathbf{a}}^{(j+1)} = -(\omega\mathbf{U} + (\omega - 1)\mathbf{D})\hat{\mathbf{a}}^{(j)} + \omega\mathbf{b} \tag{4.13}$$

in each iteration j with a positive relaxation factor ω. This procedure stops if either the maximum iteration number $N_{\mathcal{J}}$ is reached or the approximated solution converges to a predefined error bound of $||\mathbf{H}\hat{\mathbf{a}}^{(j+1)} - \mathbf{b}||^2 \leq \varepsilon_a$. Since $\hat{\mathbf{a}}^{(j)}$ represents the previous result, which is constant in the current iteration, Eq. (4.13) can be reinterpreted in a compact form as

$$\mathbf{L}_\omega \hat{\mathbf{a}}^{(j+1)} = \mathbf{b}_\omega, \tag{4.14}$$

where matrix $\mathbf{L}_\omega = (\mathbf{D} + \omega\mathbf{L})$ and vector $\mathbf{b}_\omega$ represents the right side of Eq. (4.13). Such form is exactly subject to the Forward Substitution Process. As both matrices $\hat{\mathbf{K}}_1$ and $\hat{\mathbf{K}}_2$ are block-wise diagonal, the matrix $\mathbf{L}_\omega$ also shares a similar sparse structure. Hence, calculations are performed only on a few matrix elements. Leveraging this sparsity property along with the Forward Substitution, equation set (4.12) can be solved efficiently.

4.2.2 Temporal Reliability Estimation

According to the updating procedure (2.31), it is known that the naive KCF tracker is updated in a linear interpolation fashion and solely depends on the positive learning rate β. For a not well-selected learning rate, e.g., with a big value, the updated tracker model can heavily rely on the quality of the sample extracted from current image. As the behavior of an object is difficult to anticipate, the target appearance can encounter great changes, e.g., caused by illumination variation, false classification or unclear vision condition resulted by adverse weathers (Fig. 4.1 (c)-(d)). Even if these appearance changes may only happen in a very short period, the learned appearance model could become inconsistent with the real target if these corrupt image samples are included in the training dataset (which cannot be avoided in most of current tracking works). This could further result in drifts of predicted object locations and even tracking failures.

Besides, the fixed learning rate β is insufficient to handle the trade-off between training samples from different frames. For instance, in tracking static objects, old image samples should not be quickly excluded from the training set, because they can also provide valuable information in reidentifying the target in current frame, particularly when the vision condition is recovered. However, in tracking objects with rapid deformation or rotation, the recent samples should exert more influence on the model training so that it fits the current object appearance better. Thus, a preferred solution for above problems is the dynamic weighting of historical training samples.

Inspired by the tracking works introduced in Section 4, a joint learning fashion is also employed here, so that learning the tracker model and evaluating sample reliability can be conducted concurrently. For simplicity yet without loss of generality, only the case of one expert is discussed here, i.e., to modify the objective function in (4.9) by $\hat{\alpha} = \hat{\alpha}_e$ with $e = 1$. Thus, the problem (4.9) can be rewritten as minimizing following loss function

$$\mathcal{J}(\hat{\alpha}, \mathbf{x}_t) = \|\text{diag}(\hat{\mathbf{k}}^{\mathbf{x}_t\mathbf{x}_t})\,\hat{\alpha} - \hat{\mathbf{y}}\|^2 + \lambda\,\hat{\alpha}^H \text{diag}(\hat{\mathbf{k}}^{\mathbf{x}_t\mathbf{x}_t})\hat{\alpha} \tag{4.15}$$

with respect to image sample $\mathbf{x}_t$ at frame t. Here we introduce weights for samples in a time interval of T frames. Thus, the minimization of loss function (4.15) can be extended to a joint optimization problem of

$$\begin{aligned} &\arg\min_{\boldsymbol{\theta}, \hat{\alpha}} \mathcal{L}(\boldsymbol{\theta}, \hat{\alpha}) \\ &= \arg\min_{\boldsymbol{\theta}, \hat{\alpha}} \sum_{t=1}^{T} \theta_t\, \mathcal{J}(\hat{\alpha}, \mathbf{x}_t) + \sum_{t=1}^{T} \frac{\theta_t^2}{p_t} \end{aligned} \tag{4.16}$$

subject to

$$\sum_{t=1}^{T} \theta_t = 1 \;\; \wedge \;\; \theta_t \geq 0, \; \forall t \in [1, T], \tag{4.17}$$

where $\mathcal{L}$ is the joint loss in terms of coefficient vector $\hat{\alpha}$ and weight vector $\boldsymbol{\theta} = [\theta_1, \ldots, \theta_T]$. Term $\mathbf{p} = [p_1, \ldots, p_T]$ is a regularization vector consisting of positive priors for sample weights. Constraint (4.17) implies that sample weights should be assigned with non-negative values and their sum equals 1.

Since the regularization vector $\mathbf{p}$ greatly influences the distribution of sample weights, it should be chosen reasonably. In this approach, it resorts to the memory model of human brain [5], in which the memory follows the paradigm that the recently captured objects should be always preserved, because they are with high probability to appear again, while the old ones without presence are not so important and can be gradually forgotten. This decay of memory can further reflected by an exponential function, which is called as the forgetting curve [45]. In light of these models, the prior p_t for a sample $\mathbf{x}_t$ can be initialized in a similar way, formulated as

$$p_t = \mu \exp(-\frac{T - t}{Th}), \; \forall t \in [1, T], \tag{4.18}$$

where parameter h is a positive number to control the strength of memory. The multiplier μ is a normalization factor to guarantee that the sum of all priors is normalized to 1. Since the exponential function presents the decline of memory retention over time, the strongest retention is always obtained at the current time stamp T.

Introducing Eq. (4.18) into loss function $\mathcal{L}$ in (4.16), it can be easily verified that for any fixed $\boldsymbol{\theta}$ or $\hat{\boldsymbol{\alpha}}$, the remained optimization problem is convex and thus the joint problem (4.16) is demonstrated as biconvex. Leveraging this property, the approach of *Alternate Convex Search (ACS)* [57] can be employed to solve the joint optimization problem. In this method, it iteratively solves two subproblems either with fixed $\boldsymbol{\theta}$ or $\hat{\boldsymbol{\alpha}}$. Thus, following two steps are proceeded in each iteration:

a) Updating the coefficient vector $\hat{\boldsymbol{\alpha}}$. In the beginning, the weight vector $\boldsymbol{\theta}$ is initialized with as equally distributed. In the first updating step of each iteration l, sample weights are fixed. By ignoring the last term in (4.16), the subproblem can be interpreted as minimizing following loss

$$\mathcal{L}(\hat{\boldsymbol{\alpha}}^{(l)}) = \sum_{t=1}^{T} \theta_t \, \mathcal{J}(\hat{\boldsymbol{\alpha}}^{(l)}, \mathbf{x}_t) \tag{4.19}$$

with respect to coefficient vector $\hat{\boldsymbol{\alpha}}^{(l)}$. Since this simplified form can be regarded as summing weighted loss function $\mathcal{J}(\hat{\boldsymbol{\alpha}}^{(l)}, \mathbf{x}_t)$ for each sample $\mathbf{x}_t$ with a shared coefficient vector $\hat{\boldsymbol{\alpha}}^{(l)}$, it can be solved by the approach (4.11)-(4.12) yet with following modifications

$$\begin{cases} \mathbf{H} = \sum_{t=1}^{T} \theta_t (\hat{\mathbf{K}}_{1,t}^H \hat{\mathbf{K}}_{1,t} + \lambda \hat{\mathbf{K}}_{2,t}) & \text{(4.20a)} \\ \mathbf{b} = \sum_{t=1}^{T} \theta_t \hat{\mathbf{K}}_{1,t}^H \hat{\mathbf{y}} & \text{(4.20b)} \end{cases},$$

where subscript $_t$ denotes that both matrices $\hat{\mathbf{K}}_{1,t}$ and $\hat{\mathbf{K}}_{2,t}$ are calculated by the same sample $\mathbf{x}_t$ at time t. The acquired solution $\hat{\boldsymbol{\alpha}}^{(l)}$ is then passed on to the next step to update sample weight vector $\boldsymbol{\theta}^{(l)}$.

b) Updating the weight vector $\boldsymbol{\theta}$. In the second update step, given a coefficient vector $\hat{\boldsymbol{\alpha}} = \hat{\boldsymbol{\alpha}}^{(l)}$, the loss $\mathcal{J}(\hat{\boldsymbol{\alpha}}^{(l)}, \mathbf{x}_t)$ for each individual sample $\mathbf{x}_t$ is constant. Thus, the remaining optimization problem of (4.16) only depends on the weight vector $\boldsymbol{\theta}^{(l)}$, and parameters of the joint loss $\mathcal{L}(\boldsymbol{\theta}^{(l)}, \hat{\boldsymbol{\alpha}}^{(l)})$ can then be reduced to $\mathcal{L}(\boldsymbol{\theta}^{(l)})$. Since this simplified function is formulated in a quadratic form of the weight vector $\boldsymbol{\theta}^{(l)}$ subject to constraint (4.17), it can be solved by the Quadratic Programming method, which is integrated in standard off-the-shelf solvers such as [51]. The solution $\boldsymbol{\theta}^{(l)}$ is reused for updating coefficient vector $\hat{\alpha}^{(l+1)}$ in the next iteration $l+1$.

Above procedure terminates if either the maximum iteration $N_{\mathcal{L}}$ is reached or the joint loss $\mathcal{L}$ converges to a predefined error bound, interpreted as

$$\|\mathcal{L}(\boldsymbol{\theta}^{(l)}, \hat{\alpha}^{(l)})\|_2 \leq \varepsilon_{\mathcal{L}}, \tag{4.21}$$

where $\varepsilon_{\mathcal{L}}$ denotes the predefined upper limit of approximation error. After training, samples are sorted according to their weights in an ascending order. If the amount of training dataset exceeds the predefined upper limit T, samples with the smallest weights are eliminated, so that both the computational burden and the memory consumption are bounded. The current appearance model $\mathbf{x}$ utilized in the evaluation function can be updated by aggregating weighted samples, expressed as

$$\mathbf{x} = \sum_{t=1}^{T} \theta_t \mathbf{x}_t. \tag{4.22}$$

For a qualitative impression of the whole learning procedure of introduced tracking approach, an example is illustrated in Fig. 4.4. Here $T = 60$ historical samples are kept. From this example, it can be seen that the calculated weights exhibit a similar trend with the prior curve except at frame 72. In these frames the vision is unclear due to raindrops on the windshield. The sample is corrupted and thus assigned with a very small weight. After raindrops are cleaned and the vision condition is recovered, high weight values are assigned to the image sample, e.g., at frame 79. Thus, the classifier is only trained on the most confidential samples, which improves the robustness of tracking, especially under adverse weather conditions.

4.3 Evaluation

In this section, the proposed tracking approach is evaluated in comparison with several state-of-the-art trackers which are based on the night traffic dataset [25].

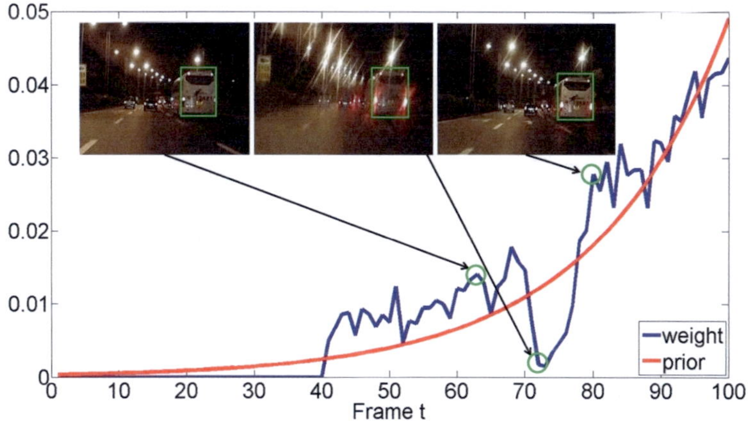

Figure 4.4: In this example, a bus is tracked in a wet night. Here $T = 60$ historical samples are kept to train the classifier. Sample weights and the priors are denoted in blue and red respectively. Since samples at frames 63 and 79 are extracted from a clear vision, they are assigned with big weights. For the sample at frame 72, because the vision is contaminated by raindrops on the windshield, target appearance becomes inconsistent with the real one. Thus, the related weight is set very low.

Firstly the robustness of proposed tracker is verified in dealing with general low illumination conditions. Thereafter, its effectiveness is investigated in handling tracking scenarios with specific attributes including the adverse weather conditions, to further prove that it can couple with corrupted image samples. In an additional runtime analysis, we show that the proposed approach also enjoys a favorable real time performance.

4.3.1 Experimental Setup

As previously mentioned in Section 4.1, there is rare systematical investigation about object tracking under deteriorated vision condition and the same problem is also confronted by most standard datasets. The most related dataset is from the work of [25], which focuses on recognition and tracking of vehicles under low illuminated condition, especially during nighttime. This dataset is chosen for evaluation because this thesis is also focused on object tracking in traffic scenarios and the low illumination condition definitely fits the motivation of the development of the proposed tracking approach. In this dataset, images are recorded by a wide range camera mounted directly behind the frontal windshield of a moving vehicle at night. Totally, it consists of 119 videos with a total length of about 3.5 hours and images are with a resolution of 856×480 pixels. Although this dataset is named with the term "night", its included sequences are actually recorded with different time slots (e.g., from nightfall to the morning), road conditions (e.g., from highways to crowded streets) and weather factors (e.g., from clear weathers to the ones with raindrops or fogs), which also helps the evaluation of proposed approach in dealing with corrupted samples. Aside from that, a lot of vehicles with a rich diversity in color, size, class and behavior are labeled in the image, which enables the evaluation of tracking in various challenging scenarios.

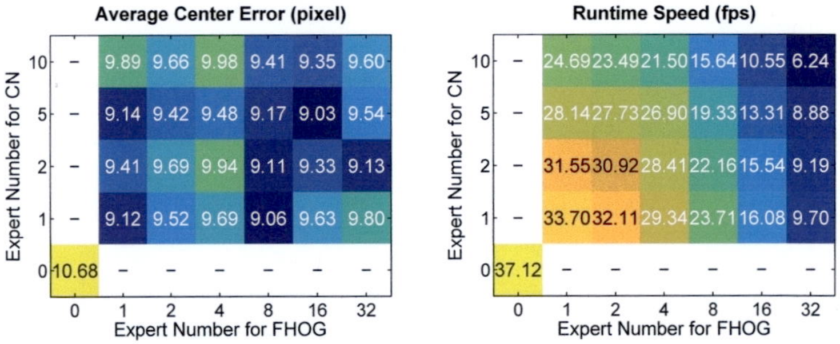

(a) (b)

Figure 4.5: Heatmaps (a)-(b) respectively show the tracking accuracy and runtime speed of proposed approach in dependence of the expert number and evaluated on dataset [25]. The first metric is measured by the average center distance between estimated target and its groundtruth, while the second one is given in frames per second. Since the channel number of FHOG [49] and CN features [40] respectively equal 32 and 10, only the expert number which can equally divide the channel number is chosen for each feature type. The evaluated tracker is KCF_CR because the temporal reliability estimation is not related here. The tracker is learned with $N_{\mathcal{L}} = 4$ iterations, which is proven to be sufficient in Section 4.3.2. The value at zero coordinates corresponds to the naive KCF tracker.

As different reliability estimations are incorporated in the proposed approach, for a better reveal of their performance, three versions of KCF tracker are prepared. The first one is the naive KCF model introduced by [71]. The next one is named as KCF_CR, which integrates aggregated experts for channel-wise reliability estimation. The last one is a combination of both channel-wise and temporal reliability estimation and is dubbed as KCF_CTR. A common appearance model is shared by above trackers and is built with the FHOG [49] and the CN features [40], which is same as the setup in last chapter. Based on a grid search method (Fig. 4.5), two different experts are utilized, with each encoding one feature type, because the deployment of more experts can only improve the tracking accuracy minorly but significantly decrease the runtime speed, as illustrated in Fig. 4.5. Empirically, the relaxation factor is set to $\omega = 1.1$. For the temporal reliability estimation, the maximum number of observed samples is equal to $T = 60$. The memory strength equals $h = 0.5$. The upper bound of approximation error ε_{α} and $\varepsilon_{\mathcal{L}}$ are equal to 10^{-4}. Other parameters of KCF-based tracker are consistent with the default setting in [71]. All experiments are preformed on a laptop platform which is with an Intel i7-3740QM CPU of 2.7GHz and a memory of 8GB. The proposed approach is implemented in C++ programs and runs only in a single thread.

4.3.2 Ablation Study on Iteration Procedures

Up to now, there are still two parameters remaining unsettled, i.e., the iteration number $N_{\mathcal{J}}$ and $N_{\mathcal{L}}$, which respectively control the optimization procedures for learning the channel-wise and temporal reliability. Since they strongly influence the performance of proposed tracker, their values had better be determined by experiments. Here the evaluation exactly follows the OPE-protocol [167] and the tracking performance is measured by two metrics: the Average Euclidean Distance (AED) and the Success Rate. The first one calculates the distance between centers of the target and its groundtruth. The upper limit is set to 20 pixels (as recommend in [167]), so that the negative impact from outliers caused by loss of the target can be reduced. The second one accumulates the area under ROC-curve with respect to a specific threshold of overlap between estimated target and its groundtruth, which equals 0.5 in this experiment. For each of those two parameters, experiments are conducted within a small range of $1 \leq N_{\mathcal{J}}, N_{\mathcal{L}} \leq 8$, so that the approximation error bound will not be reached.

The evaluated tracker is KCF_CTR but the temporal reliability estimation is deactivated (which is thus equivalent to KCF_CR) in the first experiment so that it imposes no influence on investigating the iteration number $N_{\mathcal{J}}$. The corresponding evaluation results are reported in Fig. 4.6 (a). It can be seen that the AED curve for temporal reliability estimation already converges after the 2nd iteration while its success rate only becomes stable when reaching the point of $N_{\mathcal{J}} = 3$. Since more iterations bring mere benefit for the tracking precision but can only aggregate the computational burden, the value of $N_{\mathcal{J}}$ is set to 3. In the next experiment, the temporal reliability estimation is reactivated for KCF_CTR. With fixed $N_{\mathcal{J}}$ value, it only investigates the influence of iteration number $N_{\mathcal{L}}$. Accordingly, the evaluation results are plotted in Fig. 4.6 (b). It is obvious that for the channel-wise reliability estimation, the AED curve decreases significantly in the first half range while it settles down afterwards. A similar trend can be seen on the curve of success rate but in the contrary direction: it firstly increases and then slows down after the 4th iteration. With the same reason on the trade-off between tracking-precision and computational load, the optimal value for parameter $N_{\mathcal{L}}$ is chosen as 4 in further experiments.

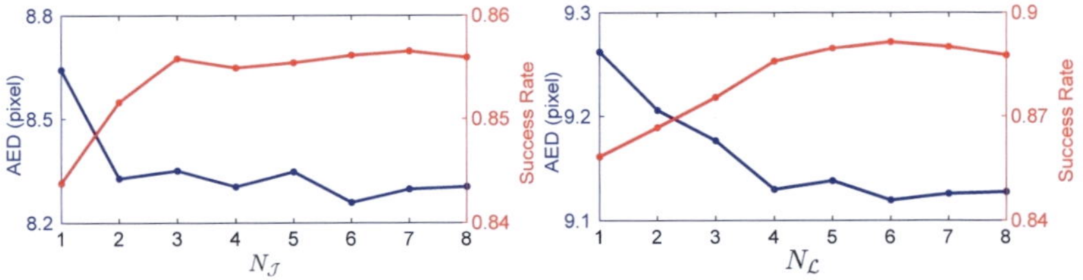

(a) (b)

Figure 4.6: Plots (a)-(b) respectively show the evaluation results of tracker KCF_CTR on varied iteration numbers of $N_{\mathcal{J}}$ and $N_{\mathcal{L}}$. The tracking precision is measured by two metrics [167]: the AED value and the success rate, which are denoted in blue and red respectively.

4.3.3 General Evaluation on Low Illumination

In the next experiment, the performance of proposed approaches, i.e., KCF_CR and KCF_CTR is compared with the baseline model KCF as well as with other seven state-of-the-art trackers, i.e., CT [40], BSBT [139], CXT [43], CSK [70], CMT [115], SPLTT [141] and SRDCFdecon [37]. Among them, the first four trackers do no employ explicit strategies to deal with vision deterioration. The BSBT adopts a boosted classifier while the others are mainly CF-based trackers and differ in utilized image features. The CMT is a point-based tracker and measures sample reliability according to the consensus of point motion while the SPLTT repeatedly verifies old samples with respect to detector responses. The SRDCFdecon is similar to the proposed tracker and joins classifier learning and sample reliability estimation into one framework. The difference is that it adopts a linear CF model and is learned on big image regions. For a fair comparison, parameters of those trackers are consistent with their original papers. Here it follows the OPE-protocol [167] to initialize the tracker at input frame with a given bounding box and thereupon record its predicted target location and size in following frames. To better explore the tracking performance, here two more metrics are employed: the temporal robustness evaluation (TRE) and the spatial robustness evaluation (SRE)[3]. In the first metric, the tracker is triggered at equally distributed time points to simulate image sequences with varied lengths. In the second metric, the tracker is only initialized at the first frame yet with a shifted or scaled bounding box, which can be considered as resulted from an imperfect object detection. The tracking performance is represented by both the precision and success plots.

Evaluation results of compared trackers are illustrated in Fig. 4.7. For each plot, trackers are ranked with respect to their precision and success rate at the threshold of a location error of 20 pixels and an overlap of 50% with the groundtruth, which is same as in the last chapter. In the ranking list it can be seen that trackers such as CT, BSBT and CXT generally exhibit a poor performance in tested cases, which can be reasoned by their deployed simple image feature, e.g., the raw pixels. As they only employ simple learning models such as CF or boosting, they cannot well handle the great degradation of visual feature in low illuminated cases. Although SPLTT repeatedly verifies its old samples, as it is only trained by a normal SVM, even if historical samples are integrated, its training set is still much smaller than those CF-based trackers, which take advantage of the circular data structure. Therefore, its performance is also unsatisfied. Since CSK employs point matching to aid the inference for target location, it achieves a relative higher precision than most trackers.

[3] In last chapter, we did not use the TRE metric, because in most tested sequences, the occlusion only occurs once. If TRE is utilized, the tracker may be frequently initialized in sub-sequences without occlusions, which strongly influences the investigation on tracker behavior in occluded cases. We did not use the SRE metric in last chapter neither, because in some cases with severe occlusion, the visible target part is less than 10%. In such cases, even a small shifting or scaling of the bounding box can lead to the risk of excluding visible object parts from the box, making the tracker initialization or learning impossible.

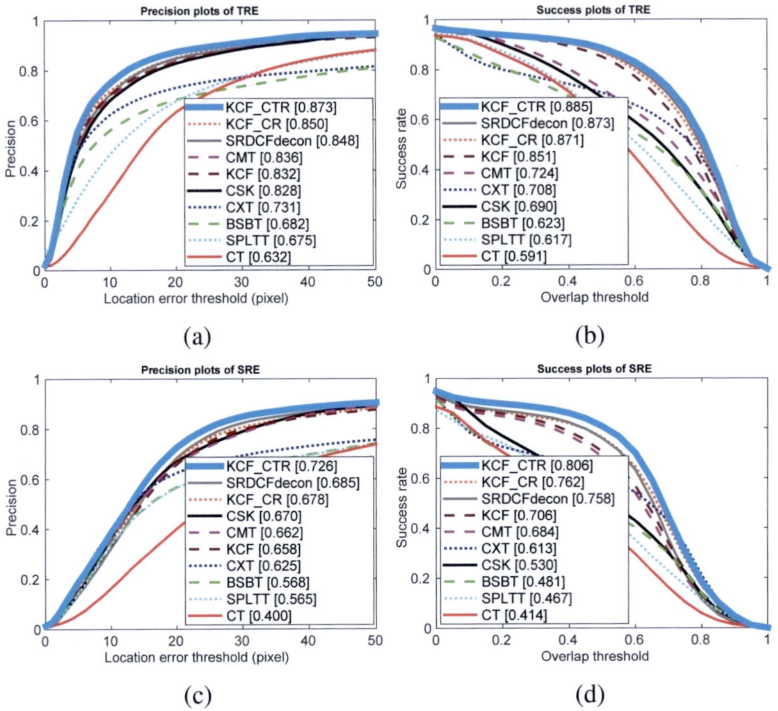

Figure 4.7: Performance evaluation on TRE and SRE metrics with each presented in one row. The first column displays precision value calculated in terms of location error while the second column displays success rate based on overlap ratio between the predicted target and its groundtruth. Proposed method is marked in bold. Additionally, trackers are ranked descendingly according to the precision value at location error of 20 pixels and success rate at an overlap ratio of 0.5.

However, points are prone to noise in low exposed images, thus the precision of its estimated object size is not high (Fig. 4.7 (b) and (d)). A similar situation can also be seen for the CMT tracker, since it also models target by dense sampled points. Due to the benefit of kernel function, the naive KCF tracker itself already outperforms most of state-of-the-art trackers. However, it still suffers from the degradation effect, which is reflected by the low precision in SRE test (Fig. 4.7 (c)). By integrating channel-wise reliability estimation, both its precision and success rate increase by about 2 ~ 5% in KCF_CR, due to a better feature weighting. Since SRDCF employs weighting in both spatial and temporal domain, it can even handle temporal varying vision deterioration. Thus, it provides a comparable or even superior performance to KCF_CR. However, it still performs inferior to the proposed KCF_CTR, which implies that the proposed joint reliability estimation over both feature channel and historical samples is more effective.

4.3.4 Attribute-based Evaluation

For a comprehensive analysis of the influence on tracking performance by different factors, five additional experiments are conducted according to attributes of color, weather, behavior, time and vehicle class. For a fair comparison, only one attribute is changed in each experiment set while the others are kept the same or equivalent. For each attributed scenario, sequences with a length of 200 to 1000 frames are chosen and the number of tracked targets is in a range of 30 to 50. The performance is measured in the same way as before. Test results are plotted in Fig. 4.8 and 4.9. A detailed description is as follows.

Color: The low illumination causes the fact that bright colors are more recognizable than the dim ones in the image. Therefore, the average tracking precision and success rate on white or red vehicles are much greater than those with gray or black colors, as displayed in Fig. 4.8 (a) and 4.9 (a). Among the tracker models, the proposed approaches KCF_CTR and KCF_CR always present a prominent performance, especially in tracking targets with strongly faded colors (Fig. 4.10 (1a)-(1c)), which demonstrates the effectiveness of feature weighting in constructing tracker models.

Weather: Here it is related to three cases: clear, rainy and foggy weather. In the second case, the vision condition is interfered by raindrops, resulting in contaminated image areas (Fig. 4.10 (2b)). Despite that the corrupt object appearance is troublesome for most trackers, by removing outliers through estimation of temporal reliability, both approaches KCF_CTR and SRDCFdecon can successfully track the object, as illustrated in Fig. 4.10 (2a)-(2c). In foggy cases, as the visible distance is mostly less than 10 meter and the distance between target and ego-car varies over time, it results in a severe vision deterioration, which is also temporally varied. However, both trackers still accurately identify the target, which further demonstrates the advantage of dynamic weighting of samples.

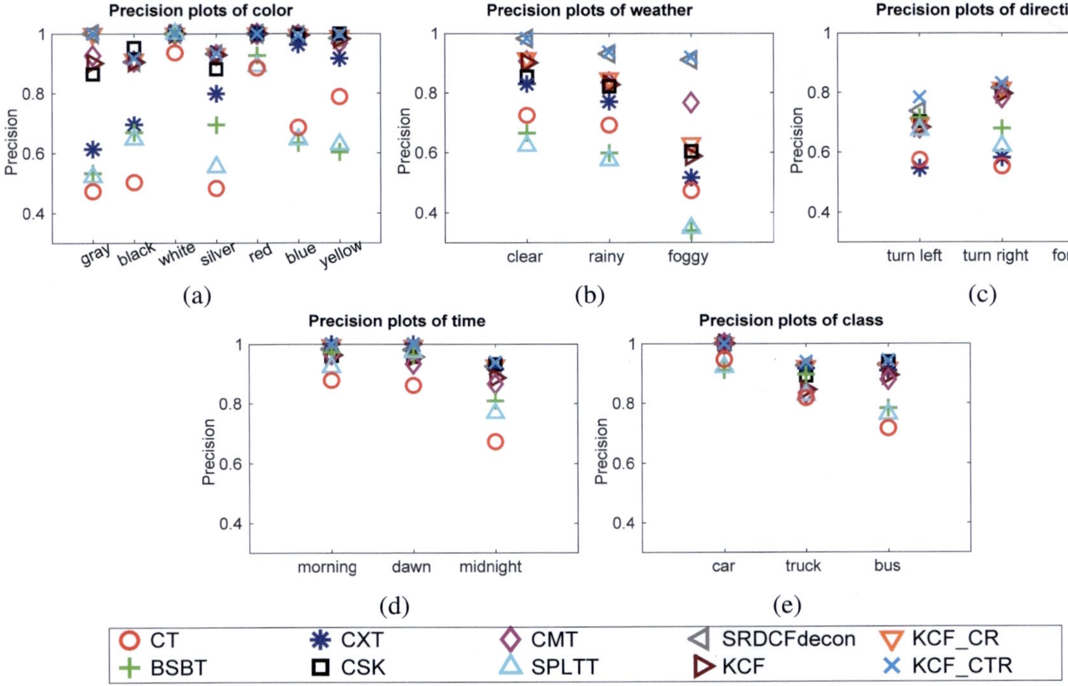

Figure 4.8: Evaluation of tracking performance in attributes of color, weather, behavior, time and vehicle class. These plots present the precision in terms of a location error of 20 pixels.

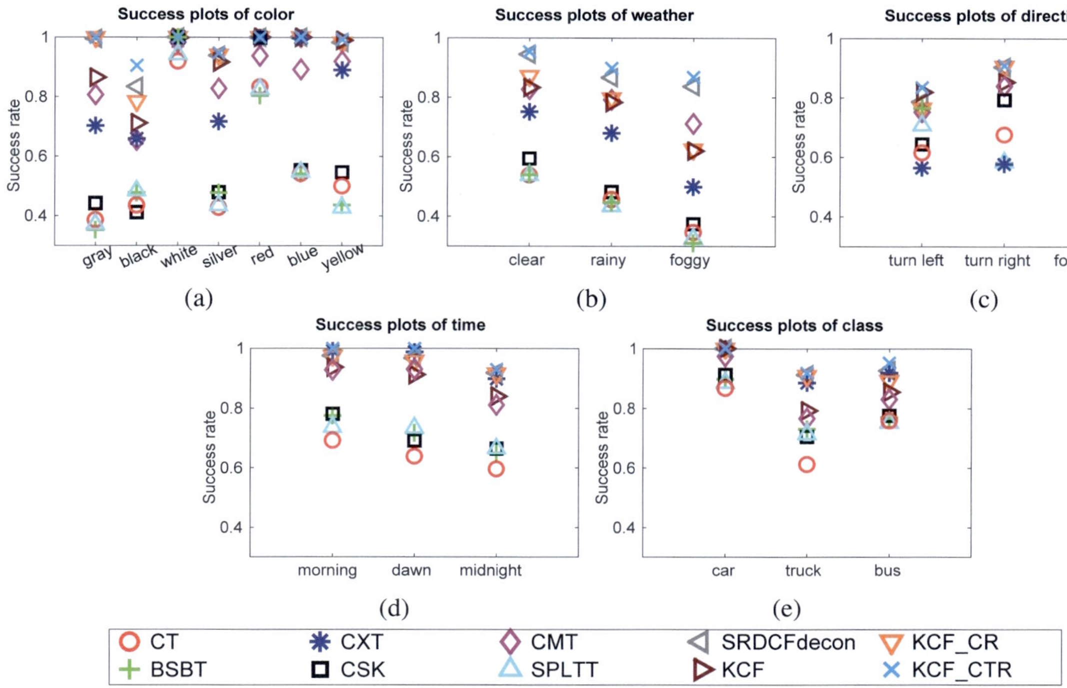

Figure 4.9: Evaluation of tracking performance in attributes of color, weather, behavior, time and vehicle class. These plots present the success rate with more than 50% of overlap between predicted target and its groundtruth.

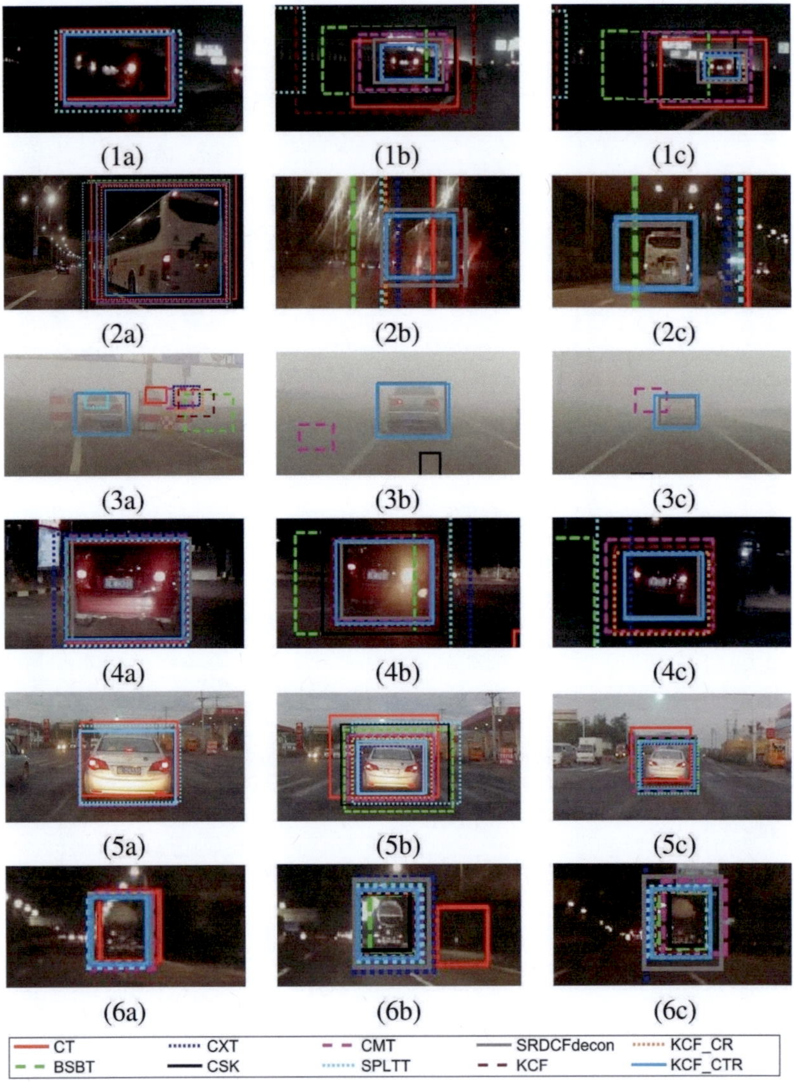

Figure 4.10: Examples of vehicle tracking in attributes of color, weather, behavior, time and vehicle class, with each row displaying one sequence.

Behavior: As depicted in Fig. 4.8 (c) and 4.9 (c), tracking turning vehicles is far more difficult than forward driving ones. The main issue is the turning signal, which disperses into bright image regions in the dark environment (Fig. 4.10 (4b)). Because the signal light blinks at a specific frequency, the dispersing effect appears at non-successive frames, leading to frequent change of object appearance. Despite that fact, it can be well handled by carefully learning the tracker according to channel-wise and temporal reliability. Hence, both KCF_CTR and KCF_CR perform well in this case.

Time: Here it focuses on three scenarios: morning, dawn and midnight. The first scenario can be regarded as test under regular daylight (Fig. 4.10 (5a)-(5c)). Since the illumination condition is relative good, the highest accuracy for all trackers are achieved. In the second scenario, captured images are a little bit gloomy. However, information like the contour, color and shape of an object is still observable. Hence, the average tracking precision obtained is higher than in midnight, where the visual information is strongly weakened by the darkness. Nevertheless, the KCF_CRT tracker can well handle all these conditions. Thus, it ranks on the top in all three cases.

Vehicle Class: Evaluation results in Fig. 4.8 (e) and 4.9 (e) imply that tracking trucks and buses are more difficult than cars under low illumination condition. The reason is that buses are usually painted with various figures and logos, e.g., ads, on different sides. This property causes great appearance change particularly in passing-by scenario, which is hard for most trackers to deal with. For trucks, the trouble mainly comes from its rear part. If the distance between ego-vehicle and tracked truck varies over time, reflection stripes of the truck may not always be reached by the frontal light of ego-vehicle, thus resulting in changed image pattern (Fig. 4.10 (6a)-(6c)). In spite of both difficult cases, the proposed KCF_CRT still shows strong robustness, even with a more accurate size estimation than the SRDCFdecon.

4.3.5 Runtime Performance Analysis

The runtime performance of tested trackers is reported in Table 4.1. On the top is the CSK approach with a speed of almost 110 fps, owing to its adopted simple model and feature, yet at the cost of proneness to the noise in low exposed images. After it follows the naive KCF tracker yet with a large gap of more than 70 fps. By integration of channel-wise weighting, the speed of KCF_CR decreases to about 34 fps, but still faster than most trackers. By further employing temporal reliability estimation, the KCF_CTR runs at 21.4 fps. Such speed is slower than the CT tracker but much faster than the rest approaches, especially the ones without utilization of circular data structure, e.g., the BSBT, SPLTT and SRDCFdecon, which are slower than 3 fps. Considering the performance gain on tracking precision, the proposed KCF_CTR tracker shows strong robustness against various vision deterioration condition. And its speed, i.e., 21.4 fps, can still fulfill most real time requirements.

Method	CT	BSBT	CXT	CSK	CMT	SPLTT	SRDCFdecon	KCF	**KCF_CR**	**KCF_CTR**
FPS	28.0	1.6	9.1	109.3	11.8	0.5	2.2	37.1	33.7	21.4

Table 4.1: Evaluation of runtime performance in fps with proposed approaches marked in bold.

5 Tracking with Multi-Object Reidentification

In previous chapters, we already talked about visual tracking under complex circumstances such as severe occlusion and vision deterioration. However, all of these scenarios discussed up to now are still related to tracking of a single object. Thus, there is one question still remained as open, i.e., the target reidentification for tracking of multiple objects.

To deal with the problem of vision-based multi-object tracking (MOT), one of the most popular trends is the tracking-by-detection paradigm, which leverages modern detection technologies [21] and is adopted in most of recent successful tracking works. In such paradigm, the inputs are only object hypotheses provided by a detector which is evaluated on each frame of a video sequence. During the tracking phase, the candidates, which are most similar to the tracked target are linked together to form its trajectories, i.e., the tracks. Due to significant progress in the domain of object detection, recent detectors can achieve an accuracy which is higher even than human beings [121]. This leads to the fact that given a well-performed detector, the precision of current tracking approach is mainly dependent on the utilized association technique. Thus, the multi-object tracking can be cast as solving an association problem or a set of subordinate ones.

To link detections with the target, most of current tracking methods prefer an enhanced similarity measurement by incorporating appearance features along with the location and size information, and pair the ones with the highest similarity [145, 147]. Although this idea can well solve the association in a small region or in a short period, it is insufficient to recover long-term lost targets. For instance, the current location of a target, which is reidentified after a long time full occlusion, may be far away from where it disappears. This is also an open question left by previous chapters but cannot be handled by simple similarity measurement strategies. Another problem is related to dynamic cases, especially when cameras are installed on moving platforms like automobiles. In these cases, tracks can be easily affected by the camera motion. Thus, it can yield misestimated target locations by conducting simple similarity measurements and further lead to association errors. Such a case is rarely considered in current tracking approaches.

Regarding these facts, in this chapter, a new multi-object tracking approach is presented, which is published in [104, 145, 147, 175] and won the experienced tracking level on the UA-DETRAC benchmark [159] and multi-object tracking task on the VisDrone challenge [174]. For efficiency, the association in this approach is decomposed into three stages: detections-to-tracklets, tracklets-to-tracks and tracks-to-tracks. In the first step, detection hypotheses across several frames are grouped into small tracklets mainly with respect to the similarity measurement. In the second step, the association is conducted according to the relative location and motion between tracked objects. This can be considered as a kind of spatial constraint. To deal with long time full occlusion, the tracked targets should be observed in a much bigger temporal domain. Thus, in the last step, tracks belonging to the same target are stitched together with respect to a specific matching process. This is considered as a temporal constraint. All three steps are interpreted by graph theory but implemented with different settings. Details about these steps are introduced in the theory part of this chapter, which directly follows the introduction of state-of-the-art works. In the last part, extensive experiments are presented about ablation study of key procedures as well as the general performance of proposed approach in different challenging situations. Through rich experimental results, it demonstrates that the association problem in multi-object tracking can be effectively solved by the proposed approach with joint constraints in both spatial and temporal domains, which yields a superior performance over most state-of-the-art methods. As the proposed approach runs online with a small computational load, it admits real time applications, which is shown at the end of this chapter. To avoid confusion of the readers, without explicit declaration, all tracking approaches discussed in this chapter, are multi-object tracking approaches.

5.1 State of the Art

Vision-based multi-object tracking has drawn increased attention from the research community during the last decades and most of the recent successful works follow the tracking-by-detection paradigm [110]. Depending on the manner how individual detections are processed, multi-object tracking approaches can be coarsely categorized into two groups: filtering-based approaches and batch-based ones [95]. Along with the rapid development of ADAS and automated driving systems, tracking approaches implemented on moving platforms becomes another active research area. Thus, in contrast to conventional methods, which focus on static cameras, dealing with dynamic images becomes an interesting point among recent works. Therefore, for a better impression about the multi-object tracking, related works herein are introduced in three perspectives: filtering-based methods, batch-based methods and those dealing with dynamic images.

Multi-Object Tracking with Filtering

In the first category, filtering-based tracking approaches mostly estimate the current location or size of a target based on its historical states, e.g., by leveraging the Markov theory. Since they only process a few detected objects (usually one) at a time, the key falls on how to attain high confidential detections for the target. Representatively, to reduce the influence of unreliable detector outputs such as false positives or mis-detections, Breitenstein et al. in [19] monitor the continuous confidence of detector and use it as graded observation model. Such a classifier is trained in an instance-specific fashion during runtime to distinguish between different tracking targets. Additionally, it combines with a particle filter to predict the target location and shows robustness against object interactions. In an advanced work [94], Lee et al. combine detection responses with *changing point detection* algorithm to observe abrupt or abnormal changes in tracked target states. Furthermore, they employ Lucas-Kanade Tracker (KLT) to calculate the likelihood of foreground object. Relying on predefined motion model, the location of the target is estimated by the method of Markov Chain Monte Carlo (MCMC). Although the number of tracking drifts and failures is greatly reduced by their approach, they still have difficulty in mapping identities between distant tracks. To tackle this problem, Xiang et al.

in [168] formulate the MOT as decision making in a Markov Process. Each state of a target encodes the information of appearance feature, location, size and all states are partitioned in predefined subspaces. Thus, both the presence and disappearance of targets are well handled in their method by simulating the detection and association as state transition between subspaces. However, the mapping policy between states and actions is required to be learned offline from training sequences.

Multi-Object Tracking with Batch Processing

Unlike in the first category, tracking by batch-based methods is usually solved over sequences by graph partition theory. In these approaches, individual detections from different frames are represented as graph nodes and linked by edges to indicate that they may be triggered by the same target. Each edge is also disbursed with a cost term with big value to penalize the assignment between non-similar objects. Thus, a subgroup of connected nodes with the minimum total cost is assumed with high probability to represent the same target. According to the problem interpretation, batch-based approaches can be further divided into two main subcategories [142]: disjoint-path-based approaches and subgraph-based ones. In the first subcategory, the MOT task is usually interpreted as a flow network, in which each node can only be linked with one node at next frame. And the maximal number of incoming or outgoing edges of one node is constrained to one. Hence, the association can be cast as searching a set of shortest paths over the sequence with minimized dissimilarity cost between their consisted nodes. To solve this min-cost flow, various recipes are proposed. Representatively, Berclaz et al. in [16] utilize an occupancy map to reduce false positives and reformulate the linking step as a convex optimization problem which fits the standard Linear Programming. Thus, it can be efficiently solved by the k-shortest paths algorithm. In another work [124], Pirsiavash et al. embed pre-processing steps such as non-maximum-suppression into a greedy algorithm and sequentially instantiate tracks on net-flow. By a dynamic programming approach, they achieve a near-optimal result within a linear processing time. Although these methods mentioned above are demonstrated to be efficient, they still have difficulties to handle the association between long ranged object hypotheses. In the subgraph-based formula, undirected subgraphs are created by clustering individual detections over frames. The nodes inside one subgraph are linked with each other. Hence, the association can be solved by searching a set of subgraphs with the minimum cost. For instance, Zamir et al. in [128] repeatedly solve the data association for one object at one time by Tabu-search and a reformulated version by Integer Linear Programming is adopted by Tang et al. in [142]. Although their method achieves a low number of ID switches, as this problem is NP-hard [12], their computation load is relative high. To bypass this bottleneck, Ristani et al. in [126] reinterpret the association task as a Binary Integer Programming (BIP) and approximate it with an online graph partition solution which is linear to the number of tracked targets. Although a near real-time performance is achieved, their work mainly focuses on tracking objects in still images.

Multi-Object Tracking with Dynamic Images

In the first two categories, we already introduced different solutions dedicated for the association problem at multi-object tracking. However, most of the above discussed approaches only consider the tracking problem in static scenes. As more and more vision-related devices are installed on moving platforms, dealing with dynamic cases, especially caused by camera motion, becomes a common issue for current tracking researches. A typical solution is proposed by Pellegrini et al. in [122], which observes the interaction between objects in bird view from a flying drone. The location of target can be predicted by a dynamic social behavior model, yet learned offline from training sequences. In [59], additional feature points are employed by Grabner et al. as supporters. The location of target can be estimated according to correlated motion between the feature points and the tracked target, so that the tracking drift resulted from moving cameras can be greatly reduced. Other than that, Yoon et al. in [170] observe the relative motion only between tracked target and adopt such structural spatial information to deal with small camera motion. Their work is further developed in [169] with spatial information integrated in an event aggregation step to search the most probable assignments. By doing this, their approach is demonstrated robust against abrupt camera motion and association ambiguities by mis-detections. However, as the object behavior is hard to predict in the image, its motion pattern can significantly differ in short and long period. Thus, to associate detections with long time lost targets can still be problematic for these methods. In contrast to them, the to be presented method aims at a tracking framework which is robust against both camera motion and association ambiguity between long ranged object hypotheses. To tackle this problem, joint constraint is utilized in terms of both spatial and temporal domains, which is introduced in following section.

5.2 Tracking by Joint Constraints

The association in both spatial and temporal domain usually requires to calculate the similarity between each two objects, which can be computational expensive or even intractable for large number of objects. Therefore, the proposed approach adopts a divide-and-conquer strategy to decompose the entire association in three separate stages. Since all of them rely on graph theory, in the beginning of this section, a brief review is given about the subgraph-based formula. Thereafter, we respectively introduce each stage, i.e., detections-to-tracklets, tracklets-to-tracks and tracks-to-tracks, and explain their benefits on association tasks. The last two stages are respectively called as spatial and temporal constraint. In the experimental part, it demonstrates that by employing such divide-and-conquer strategy, the processing efficiency can be greatly improved while satisfying result can still be obtained.

5.2.1 The Subgraph-based Formula

Leveraging the graph theory, in the MOT problem, the relationship between tracked targets and their observations (i.e., detection hypotheses) can be interpreted as an undirected graph $G = (\boldsymbol{V}, \boldsymbol{E}, \boldsymbol{\Omega})$, where $\boldsymbol{V}$ is the set of nodes standing for individual detections and their similarities are represented by the edge set $\boldsymbol{E}$. The set $\boldsymbol{\Omega} : \boldsymbol{E} \rightarrow \mathbb{R}$, contains cost for each similarity measurement, with great value to penalize the assignment between non-similar object hypotheses. The assumption utilized here is that each target in one frame can only trigger at most one shot of the detector. This can be achieved by applying non-maximum-suppression over raw detection results. Thereby, the edge set can be defined as $\boldsymbol{E} = \{(u^i, v^j) \mid i \neq j \text{ and } u, v \in \boldsymbol{V}\}$, where superscript i and j respectively denote the frame IDs for detection u and v.

In this case, the association problem can be solved by partitioning the graph G into a set of subgraphs $G_s = (\boldsymbol{V}_s, \boldsymbol{E}_s, \boldsymbol{\Omega}_s)$, where each only consists of detections with common target identity and can be interpreted by $\boldsymbol{E}_s = \{(p,q) \mid \forall p,q \in \boldsymbol{V}_s\}$ and $\boldsymbol{\Omega}_s = \{\boldsymbol{\Omega}(p,q) \mid \forall p,q \in \boldsymbol{V}_s\}$. As in each subgraph, its nodes are fully connected (i.e., each node is connected with all the other nodes), it is called as clique. The problem of partitioning an undirected graph into cliques can be reformulated as optimizing the following objective function

$$\arg\min_{\boldsymbol{X}} \sum_{(u,v)\in\boldsymbol{E}} \boldsymbol{\Omega}(u,v) \cdot x_{uv} \tag{5.1}$$

subject to

$$\begin{cases} x_{uv} \in \{0,1\},\ \forall(u,v) \in \boldsymbol{E} & \text{(5.2a)} \\ x_{uv} + x_{vw} \leq 1 + x_{uw},\ \forall(u,v),(v,w),(u,w) \in \boldsymbol{E} & \text{(5.2b)} \end{cases}.$$

Here the indicator x_{uv} is a binary encoder of the edge $\boldsymbol{E}(u,v)$ and it is set to 1 if both detections u and v share the same target identity. In constraint (5.2b)[1], the equal symbol is only then valid, when all detections u, v and w are co-identical, which means that their corresponding nodes belong to the same clique.

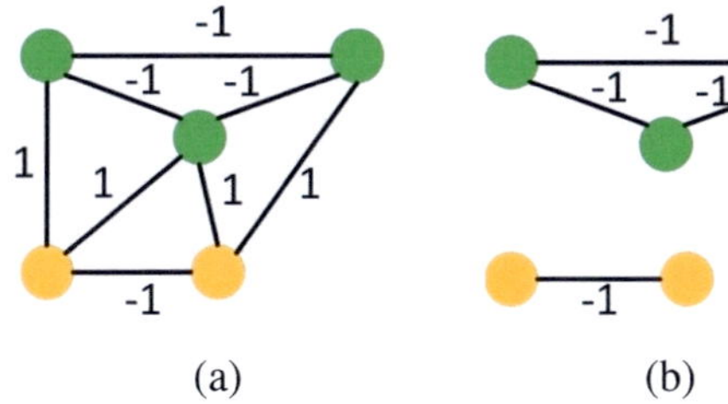

Figure 5.1: Image (a) displays an undirected graph with its nodes representing individual detections. Its edges stand for their similarities, with assigned values to indicate the costs. Image (b) depicts the solution for graph partitioning which consists of two subgraphs with co-identical nodes displayed in the same color.

[1] Such constraint also rejects structures such as rings with a length bigger than 3. Since in the ring structure, it only requires the similarity between successive nodes, different objects can be linked in a ring by nodes with ambiguous feature vectors (e.g., due to imperfect detections of crowded pedestrians). However, in a clique, such case can be avoided, because each object/node is required to be similar to all the other ones.

A feasible solution to problem (5.1) can be interpreted as a set $\boldsymbol{X}$ consisting of all edge indicators with those for co-identical node pairs set to 1 (see Fig. 5.1). As only binary values are considered in the objective function, it can be solved by the BIP algorithm, e.g., with standard solvers introduced in [126, 128].

5.2.2 Tracklet Creation

Instead of direct conducting association between individual detection hypotheses, here the MOT is solved based on tracklets due to two reasons. On the one hand, the linked detections in the same tracklet are of high similarities between each other, indicating a high probability that they describe the same target. Thus, further changes to their connections should be rare and more efforts can be spent on dealing with difficult association ambiguities. In the meanwhile, since the searching space for data association shrinks, the computational burden can be greatly reduced. On the other hand, a tracklet can be considered as a set of detections extended over frames, which enables us to extract motion information to enhance the similarity measurement. Although the target may suffer from significant motion change in long time observation, its motion in a short period changes minor and thus can be approximated by a simple model, e.g., with constant velocity. This information is helpful especially in cases, where the object matching procedure by appearance feature alone cannot give a clear solultion.

For creating tracklets, video sequence are first divided into small segments with each consisting of f_l consecutive frames. Each detection is represented by a parameter tuple $D = (\boldsymbol{\varphi}, \mathbf{b}, t, \mathbf{v})$, where term $\boldsymbol{\varphi}$ is a feature vector describing object appearance and vector $\mathbf{b} = [p_x, p_y, w, h]^\top$ denotes the bounding box including the center location $\mathbf{p} = [p_x, p_y]^\top$ and the size $\mathbf{z} = [w, h]^\top$. The frame ID is indicated by t and vector $\mathbf{v} = [v_x, v_y]^\top$ denotes the 2-D velocity. For two detections $D_1 = (\boldsymbol{\varphi}_1, \mathbf{b}_1, t_1, \mathbf{v}_1)$ and $D_2 = (\boldsymbol{\varphi}_2, \mathbf{b}_2, t_2, \mathbf{v}_2)$, the appearance affinity is defined as

$$s_a = \mathrm{corr}(\boldsymbol{\varphi}_1, \boldsymbol{\varphi}_2) \tag{5.3}$$

by computing the correlation coefficient between their appearance descriptors. Moreover, their size affinity is defined as

$$s_z = 1 - \left\| \frac{\mathbf{z}_1 - \mathbf{z}_2}{\mathbf{z}_1 + \mathbf{z}_2} \right\|_2, \tag{5.4}$$

where the fraction operation denotes an element-wise division. As detections are given with different sizes, to calculate the appearance descriptor $\boldsymbol{\varphi}$, candidate image patches are first rescaled to a unified size, e.g., 64×64 pixels. After that, gradient and color features as deployed in previous chapters are extracted from the image and then concatenated into one vector. To reduce the influence of background objects especially in boundary regions, the image patch is multiplied with a 2-D Hanning filter. The assumption is that the central area within the patch has a high probability of belonging to the target, which is valid for most object classes.

In the proposed approach, it also assumes that the motion of an object in the image is limited in a small number of frames. This is also valid for moving cameras with a high frame rate. Under this assumption, the velocity of a detection hypothesis $D = (\boldsymbol{\varphi}, \mathbf{b}, t, \mathbf{v})$ can be estimated with the help of its nearest neighbor $D_k = (\boldsymbol{\varphi}_k, \mathbf{b}_k, t_k, \mathbf{v}_k)$ by

$$\mathbf{v} = \frac{\mathbf{p} - \mathbf{p}_k}{t - t_k} \tag{5.5}$$

subject to

$$\begin{cases} 0 < |t - t_k| \leq \theta_t & (5.6a) \\ \|\mathbf{p} - \mathbf{p}_k\|_2 \leq \theta_p & (5.6b) \\ s_a > \theta_a & (5.6c) \\ s_z > \theta_z & (5.6d) \end{cases},$$

where θ_t indicates the predefined small frame interval and θ_p denotes the gating radius. Constraints (5.6c)-(5.6d) represent that affinity values s_a and s_z should be greater than the predefined thresholds θ_a and θ_z. For isolated detections, they are assigned with zero velocity.

In light of above definitions, the similarity between two detections D_1 and D_2 can be calculated by

$$s_d = s_a \cdot s_z \cdot s_p \tag{5.7}$$

with

$$\begin{cases} s_p = \max\left(1 - \gamma(l_{1,2} + l_{2,1}), 0\right) & (5.8a) \\ l_{1,2} = \|(\mathbf{p}_2 + (t_1 - t_2)\mathbf{v}_2 - \mathbf{p}_1) / \mathbf{z}_1\|_2 & (5.8b) \\ l_{2,1} = \|(\mathbf{p}_1 + (t_2 - t_1)\mathbf{v}_1 - \mathbf{p}_2) / \mathbf{z}_2\|_2 & (5.8c) \end{cases},$$

where s_p denotes the position affinity. The positive balance factor γ constrains the tolerance for the forward location error $l_{1,2}$, which is defined by the distance between the location of D_1 and the estimated location of D_2 at frame t_1. Such a distance is further normalized by object size $\mathbf{z}_1$. The backward location error $l_{2,1}$ is calculated in a similar way. Afterwards, the affinity value s_d is mapped to an edge cost by

$$\boldsymbol{\Omega}(D_1, D_2) = -\ln(\lambda \cdot s_d) \tag{5.9}$$

with a small scaling factor λ. Introducing Eq. (5.9) into objective function (5.1), subgraphs, with each representing a tracklet, can be obtained by applying the BIP algorithm.

5.2.3 Spatially Constrained Association

In the image plane, a track encodes the trajectory of an object over a video sequence and its state usually incorporates the object appearance, location, size and motion information. In this regard, the track state at frame t can be denoted as $S^t = (\tilde{\boldsymbol{\varphi}}^t, \tilde{\mathbf{p}}^t, \tilde{\mathbf{z}}^t, \tilde{\mathbf{v}}^t)$ with the symbol ~ to distinguish its parameters from those of detections. Given two tracks, each with a state of $S_i^t = (\tilde{\boldsymbol{\varphi}}_i^t, \tilde{\mathbf{p}}_i^t, \tilde{\mathbf{z}}_i^t, \tilde{\mathbf{v}}_i^t)$ and $S_j^t = (\tilde{\boldsymbol{\varphi}}_j^t, \tilde{\mathbf{p}}_j^t, \tilde{\mathbf{z}}_j^t, \tilde{\mathbf{v}}_j^t)$, the spatial constraint between them can be defined by the pairwise motion pattern as $\boldsymbol{\kappa}_{i,j}^t = \boldsymbol{\kappa}(S_i^t, S_j^t) = [(\tilde{\mathbf{p}}_i^t - \tilde{\mathbf{p}}_j^t); (\tilde{\mathbf{v}}_i^t - \tilde{\mathbf{v}}_j^t)]$. Presuming that there is a number M^t of tracks at time t, the set of spatial constraints for track state S_i^t thus can be formulated as $\boldsymbol{K}_i^t = \{\boldsymbol{\kappa}_{i,j}^t | \, 1 \leq j \leq M^t\}$. Note that the case of $i = j$ is not omitted here, because the motion of object can also be estimated by its previous states, e.g., with the Kalman filter.

Relying on the spatial constraint, the location of a track i at time $t+1$ can be estimated by

$$\hat{\mathbf{p}}_{i,j}^{t+\Delta t} = \mathbf{F}\boldsymbol{\kappa}_{i,j}^t + \tilde{\mathbf{p}}_j^t \tag{5.10}$$

with respect to a transition matrix

$$\mathbf{F} = \begin{bmatrix} 1 & 0 & \Delta t & 0 \\ 0 & 1 & 0 & \Delta t \end{bmatrix}, \tag{5.11}$$

where Δt indicates the frame difference and here is equal to 1. Thus, the predicted object status can be approximated as $\hat{D}_{i,j}^{t+1} = (\tilde{\boldsymbol{\varphi}}_i^t, \hat{\mathbf{b}}_{i,j}^{t+1}, t+1, \tilde{\mathbf{v}}_i^t)$

with the predicted bounding box $\hat{\mathbf{b}}_{i,j}^{t+1} = [\,\hat{\mathbf{p}}_{i,j}^{t+1}\,;\,\tilde{\mathbf{z}}_i^t]$. Normally, to handle the association for detections-to-tracks, the predicted object should be matched with each detection at current frame to search the most similar candidate and then add it to the trajectory. Thus, in analogy to Eq. (5.7), the similarity between the predicted object $\hat{D}_{i,j}^{t+1}$ and the k-th detection D_k^{t+1} at frame $t+1$ can be measured as

$$\begin{aligned} s_d(\hat{D}_{i,j}^{t+1}, D_k^{t+1}) =& s_a(\hat{D}_{i,j}^{t+1}, D_k^{t+1}) \cdot s_z(\hat{D}_{i,j}^{t+1}, D_k^{t+1}) \\ & \cdot s_p(\hat{D}_{i,j}^{t+1}, D_k^{t+1}). \end{aligned} \tag{5.12}$$

In a further step, the similarity is mapped to a cost term as

$$\Omega_{i,j}^k = -\ln\left(\mu \cdot s_d(\hat{D}_{i,j}^{t+1}, D_k^{t+1})\right) \tag{5.13}$$

with a small scaling factor μ. Thus, the entire cost for the set of spatial constraints $\boldsymbol{K}_t^i$ can be computed as

$$\Omega_i^k = \sum_j \Omega_{i,j}^k. \tag{5.14}$$

If an object is not assigned to any detection, its cost Ω_i^0 is set to a constant value $\tau = 5$. In this manner, the object association can be cast as solving following optimization problem

$$\arg\min_{\mathbf{A}} \sum_i \sum_k \Omega_i^k \cdot a_{i,k} \tag{5.15}$$

subject to

$$\sum_{\substack{i \\ k \neq 0}} a_{i,k} \leq 1 \;\wedge\; \sum_k a_{i,k} = 1 \;\wedge\; \sum_i a_{i,0} \leq M, \tag{5.16}$$

where the solution $\mathbf{A} = \left[a_{i,k}\right]_{M\times(N+1)}$ is an assignment matrix with binary elements $a_{i,k} \in \{0,1\}$ while N and M respectively denote the number of detections and tracks. Here the frame index is omitted for generality. Constraint (5.16) indicates that all assignments are bijective except for mis-detected objects. Therefore, the association problem can be solved by the Hungarian algorithm [91]. Since the motion pattern between objects is utilized to measure their similarity, the tracking is robust against camera motion, as demonstrated in the experimental part.

Based on aforementioned procedures, the spatial constraint can be extended to deal with association for tracklets-to-tracks. Here a tracklet is denoted as a tuple $T = (\boldsymbol{D}, t^s, t^e)$, with t^s and t^e respectively to indicate its start and end frame. The set $\boldsymbol{D} = \{D^t | t^s \leq t \leq t^e\}$ consists of all its linked detections. In a similar way, here a track is denoted as $\tilde{T} = (\boldsymbol{S}, \tilde{t}^s, \tilde{t}^e)$ with its start and end frame respectively denoted as $\tilde{t}^s$ and $\tilde{t}^e$. And set $\boldsymbol{S} = \{S^t | \tilde{t}^s \leq t \leq \tilde{t}^e\}$ includes all track states at related frames. Since a tracklet extends over multiple frames, the main problem is how to match the predicted object with the tracklet. Intuitively, the predicted object can be matched with all detections in the tracklet. However, such idea brings two disadvantages. On the one hand, it extremely aggregates the computational load, which is linear to the number of detections. On the other hand, a detection in a long tracklet and from a time point which is far away from the current frame can significantly differ from the predicted object, which may lead to matching errors. Regarding these points, in this approach, following rules are proposed to deal with association for tracklets-to-tracks and illustrated in Fig. 5.2.

(i) We only check the assignment between a track $\tilde{T}$ and a tracklet T, when they do not share common frames, which is interpreted as $\tilde{t}^e < t^s$ (see Fig. 5.2 (b)). This rule coincides with the assumption that one object can only trigger at most one detection at one frame.

(ii) For efficiency, the predicted object is only matched with the first detection of a tracklet validated by rule (i). Comparing with any other detection in the tracklet, the temporal distance between the predicted object and the first detection is the shortest, thus their similarity should be high. In this case, the computational load is proportional to the tracklet number.

(iii) Based on rule (ii), we predict objects at the first frame of each tracklet. This can be attained by Eq. (5.10) yet with minor changes of $\Delta t = t^s - t_k$, where t^s denotes the start frame of tracklet T and the frame index t_k comes from track $\tilde{T}_k$, which participates in the k-th spatial constraint of T. If no common frames between them are found, the index t_k is set to the last frame $\tilde{t}^e_k$. Otherwise, the track state at frame $t_k = t^s$ is deployed for object prediction (see Fig. 5.2 (c)).

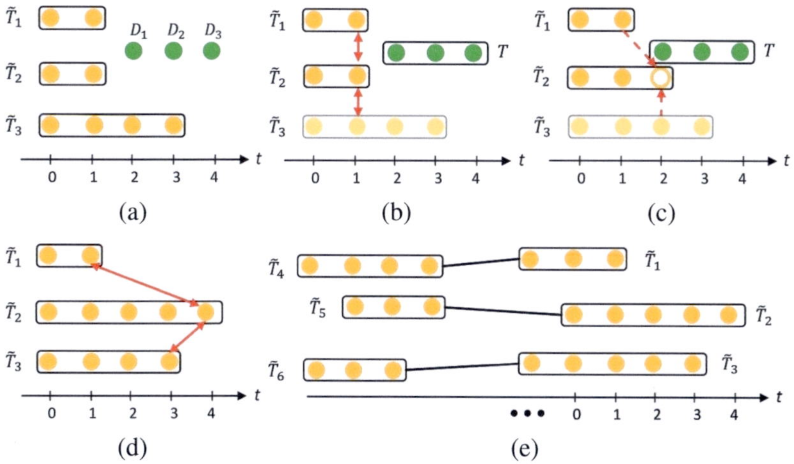

Figure 5.2: Image (a) shows three tracks $\tilde{T}_1$ to $\tilde{T}_3$ with three detections D_1 to D_3. Orange circles represent track states while detections are denoted in green. Based on affinity measurement, these detections are linked to form one tracklet T in image (b). In parallel, spatial constraints are constructed for track $\tilde{T}_2$. Each double-headed arrow links a pair of tracks, between which the motion patterns are considered. The track which is not considered in next association step is indicated by fainted color. A new object (empty circle) is then predicted for track $\tilde{T}_2$ at frame $t = 2$ in image (c). The root of single-headed arrow indicates based on which track state the prediction is made. After association, the spatial constraint for track $\tilde{T}_2$ is updated in image (d). Image (e) shows association between long ranged objects by temporal constraint.

(iv) To handle the case with only one object appearing in one frame, the spatial constraint is extended over multiple frames. For example, for each state $S_i^{t_i}$ of track $\tilde{T}_i$, its temporally nearest state $S_j^{t_j}$ from another track $\tilde{T}_j$ is searched in the time interval $|t_i - t_j| \leq m$ with the setting of $m = 3$ in the experiment (see Fig. 5.2 (d)). Thereafter, the selected states are used to construct the spatial constraint $\boldsymbol{\kappa}(S_i^{t_i}, S_j^{t_j})$.

Since a tracklet may contain non-successive frames, after association, the states in a track may not be temporally continuous. For those missing track states, their locations, sizes and velocities can be estimated by spatial constraints. In the case that more than one spatial constraint is utilized, the median value is chosen over different estimations. Appearance features for these states are approximated by a cubic interpolation based on the historical states. For efficiency, in this approach, spatial constraints are only kept for 1 second (i.e., the buffer size equals the frame rate f_r) and they are updated for each track directly after the association, which is solved by optimizing the objective function (5.15). Tracklets, which are without associations, are considered as new object tracks.

5.2.4 Long Range Association and Online Processing

Since the motion of a target may greatly change in a long period, the spatial constraint alone, which usually focuses on a short time interval, is insufficient to recover a long-term lost target, which may be temporally or spatially far away from its disappearing point. To link long ranged objects, additional strategy, i.e., the temporal constraint, should be applied on the current tracks.

This yields the new association problem: tracks-to-tracks. To solve their assignments, it can resort to the subgraph-based formula in Section 5.2.1. The main difference from previous procedures is that here the nodes of a graph represent tracks instead of individual detections. Leveraging rule (i), association is only conducted between non-overlapping tracks. For two valid tracks $\tilde{T}_1 = (\boldsymbol{S}_1, \tilde{t}_1^s, \tilde{t}_1^e)$ and $\tilde{T}_2 = (\boldsymbol{S}_2, \tilde{t}_2^s, \tilde{t}_2^e)$ with $\tilde{t}_1^e < \tilde{t}_2^s$, according to the temporal order, a possible assignment can only exist between frame $\tilde{t}_1^e$ and $\tilde{t}_2^s$. Hence, the similarity of those two tracks is defined as

$$s_d(\tilde{T}_1, \tilde{T}_2) = s_a(\tilde{\boldsymbol{\varphi}}_1^e, \tilde{\boldsymbol{\varphi}}_2^s) \cdot s_z(\tilde{\mathbf{z}}_1^e, \tilde{\mathbf{z}}_2^s) \cdot s_p(S_1^e, S_2^s), \tag{5.17}$$

where $S_1^e = (\tilde{\boldsymbol{\varphi}}_1^e, \tilde{\mathbf{p}}_1^e, \tilde{\mathbf{z}}_1^e, \tilde{\mathbf{v}}_1^e)$ indicates the state of track $\tilde{T}_1$ at its last frame $\tilde{t}_1^e$ and $S_2^s = (\tilde{\boldsymbol{\varphi}}_2^s, \tilde{\mathbf{p}}_2^s, \tilde{\mathbf{z}}_2^s, \tilde{\mathbf{v}}_2^s)$ represents the track state of $\tilde{T}_2$ at its first frame $\tilde{t}_2^s$.

Here a polynomial of second order is exploited to approximate the object location of a track $\tilde{T}_i$ at frame t, formulated as

$$\begin{cases} p_{x,i}(t) = a_{x,i} \cdot t^2 + b_{x,i} \cdot t + c_{x,i} & \text{(5.18a)} \\ p_{y,i}(t) = a_{y,i} \cdot t^2 + b_{y,i} \cdot t + c_{y,i} & \text{(5.18b)} \end{cases},$$

where $\hat{\mathbf{p}}_i(t) = [p_{x,i}(t), p_{y,i}(t)]^\top$ represents the estimated target location. And coefficient vector $[a_{x,i}, b_{x,i}, c_{x,i}, a_{y,i}, b_{y,i}, c_{y,i}]^\top$ can be attained by a regression method, i.e., the Least-Square algorithm. According to Eq. (5.5), velocity $\tilde{\mathbf{v}}_1^e$

and $\tilde{\mathbf{v}}_2^s$ in previous procedures are only estimated by detections in a short time interval. However, it is inappropriate to directly use them to calculate the location affinity s_p, particularly between long ranged tracks. Since the motion model with constant velocity may be invalid in such a long period, the affinity measurement could be erroneous. To improve the accuracy, here both the forward and backward location error $l_{1,2}$ and $l_{2,1}$ are calculated relying on the new estimated location $\hat{\mathbf{p}}_1(\tilde{t}_2^s)$ and $\hat{\mathbf{p}}_2(\tilde{t}_1^e)$. Hence, Eq. (5.8b)-(5.8c) should be modified as

$$\begin{cases} l_{1,2} = \|\left(\hat{\mathbf{p}}_2(\tilde{t}_1^e) - \tilde{\mathbf{p}}_1^e\right) / \tilde{\mathbf{z}}_1^e\|_2 & (5.19a) \\ l_{2,1} = \|\left(\hat{\mathbf{p}}_1(\tilde{t}_2^s) - \tilde{\mathbf{p}}_2^s\right) / \tilde{\mathbf{z}}_2^s\|_2 & (5.19b) \end{cases},$$

where $\hat{\mathbf{p}}_1(\cdot)$ and $\hat{\mathbf{p}}_2(\cdot)$ respectively indicate the estimation function for the track $\tilde{T}_1$ and $\tilde{T}_2$.

Regarding the application in real world, it is impossible to observe tracks over the entire time axis, which is infinitely long. Therefore, the association can only be conducted for tracks in a limited time interval, i.e., in the last f_t frames. The assumption is that much older objects can be ignored, because they have low probability to appear again. This is also consistent with the memory model of human brain [5]. For computational efficiency, the association task is performed in a batch fashion and the batch size corresponds to a time window of $f_t = N f_l$ frames. Only the tracks, which are located within this window or intersect with its boundaries, are processed. Based on measured similarities between these tracks, assignments are established according to the procedure introduced above. The batch window is then shifted forward by a constant step, which is half of its size in the experiment, and the same procedure is performed again. Here the scaling factor N is set as an even number to reduce the clipping of tracklets or tracks by a shifted window. As the association only takes place at specific frames, the average processing time falling on each frame is little.

Similar to creating tracklets, the association for tracks-to-tracks can also be solved by the BIP algorithm. However, as tracks are observed in a much larger time domain, the number of possible assignments could be huge. Solving the association problem by the naive BIP approach, which is demonstrated as NP-hard [12], may consume large memory and processing time and thus becomes difficult for applications with real time requirement. For a fast computation, here the naive BIP is approximated by the algorithm of Adaptive Label Iterative

Conditional Modes (AL-ICM) [10]. In this algorithm, the association task is formulated in similar form of objective function (5.1), and interpreted as

$$\arg\min_{\boldsymbol{L}} \sum_{(u,v)\in \boldsymbol{E}} \boldsymbol{\Omega}'(u,v) \cdot x'_{uv}, \tag{5.20}$$

where binary indicator x'_{uv} is set to 1 if two nodes are assigned with the same label. The final output is a label set $\boldsymbol{L} = \{l_1, l_2, \cdots\}$. In fact, the direct solution of problem (5.20) is a matrix $\mathbf{X}'$ consisting of binary elements x'_{uv} to indicate optimized associations, similar to the solution in Eq. (5.1). Based on this matrix, associated objects are clustered together and an unique label is assigned to each cluster, thus resulting in the final label set $\boldsymbol{L}$. In the binary matrix $\mathbf{X}'$, an indicator x'_{uv} is only then set to 1, when the node u and v are co-identical, which implies that they are assigned with the same label. The cost term for each edge in the graph is defined as

$$\boldsymbol{\Omega}'(u,v) = \ln(\lambda \cdot s_d(u,v)), \tag{5.21}$$

where the scaling factor λ is well selected so that negative cost should be assigned to a node pair with high similarity and dissimilar node pairs are penalized with positive cost. Hence, the total cost decreases when similar nodes are linked and dissimilar ones are assigned with different labels. Since the method AL-ICM is able to be scaled to large problems with a relative low computation load, as reported in [126], a fast processing speed can be achieved.

5.3 Evaluation

In this section, the performance of the proposed tracking framework is evaluated on publicly available video sequences in comparison with other state-of-the-art approaches. Here experiments are conducted on three challenging benchmarks: KITTI [95], MOT16 [110] and UA-DETRAC [159]. On the first two datasets, the effectiveness of the proposed approach is verified on tracking objects in scenarios with varied dynamics and platforms. The last dataset is involved with the impact of environmental illumination and adverse weather factors. Thus, we evaluate the robustness of the proposed approach as well as its performance when synthesized with trackers introduced in previous chapters. Since the proposed approach adopts batch process for computational efficiency, its impact on tracking performance is also explored, which is reported in the beginning of the experimental part. Through an additional runtime analysis, it shows that the proposed approach can permit a real time speed in most use cases.

5.3.1 Experimental Setup

Since this thesis focuses on object tracking in traffic scenarios and the main topic of this chapter is about reidentification of multi-objects, the most related datasets for evaluation are the benchmarks of KITTI [95], MOT16 [110] and UA-DETRAC [159]. The dataset of MOT16 consists of video sequences filmed with cameras installed on varied platforms from a small trolley to trams. It mainly observes behavior of crowded pedestrians yet with different viewing angles. In the UA-DETRAC dataset, the camera is installed on high infrastructures like poles of traffic lights or street lamps to monitor the traffic flow, especially the vehicles, yet under different illumination and weather conditions. In the KITTI dataset, cars, pedestrians and cyclists are all observed in traffic scenarios. The images are taken from a camera which is mounted on a moving vehicle for test cases with varied dynamics. In all three benchmarks there are totally about 27000 images with a resolution in the range of $0.3 \sim 2$ mega pixels. All benchmarks provide datasets for both training and test purpose. Reference detections are also given for the a fair comparison with different approaches.

In experiments, the evaluation follows the CLEAR MOT protocol [79] with six main metrics: the Multiple Object Tracking Accuracy (MOTA), the Multiple Object Tracking Precision (MOTP), the ratio of mostly tracked (MT) and mostly lost targets (ML), the number of identity switches (IDS) and the number of track fragments (FR). The MOTA value takes the number of false positives $F_p(t)$, false negatives $F_n(t)$ and identity switches $\mathrm{IDS}(t)$ as tracking errors and is interpreted as $\mathrm{MOTA} = 1 - \sum_t \left(F_p(t) + F_n(t) + \mathrm{IDS}(t)\right) / \sum_t \mathrm{GT}(t)$, where $\mathrm{GT}(t)$ denotes the ground-truth at time t. The MOTP value is measured by the overlap ratio between the estimated object and its groundtruth. These two metrics (especially the first one) are employed in the vast majority of research works to evaluate the performance of tracking approaches. Detailed description about above metrics is provided in Table 5.1.

Metric	Description
MOTA ↑	The accuracy in terms of false positives, missed targets and identity switches.
MOTP ↑	The alignment between matched GT and predicted bounding boxes.
MT ↑	The ratio of groundtruth tracks that are covered by a hypothesis for at least 80%.
ML ↓	The ratio of groundtruth tracks that are covered by a hypothesis for at most 20%.
IDS ↓	The total number that an object switches its matched GT identity.
FR ↓	The total number of times a GT trajectory is fragmented/interrupted.

Table 5.1: Description about evaluation metrics on tracking performance [159]. The symbol of an up-arrow means that higher values are preferred. And a down-arrow implies that lower values are better.

Since in real cases like ADAS and automated driving systems the data is mostly processed online, to fulfill such requirement, an online mode is also employed for the proposed approach, i.e., in experiments the association is forced to be conducted only on the frames and detections up to the current frame. For parameter settings[2], in the proposed approach, the velocity of each detection hypothesis is estimated in a neighborhood of $\theta_t = 3$ frames. The gating radius θ_p is determined by the larger dimension of the object size, which is scaled by a factor of 2. It empirically sets the parameters $\lambda = 5$ and $\mu = 1$. The thresholds θ_a and θ_z are respectively set to 0.5 and 0.3. The balancing factor γ is equal to 0.5. As the association is always done batch-wisely, if the sequence length is not an integer number of it, there will be unprocessed frames at the end of the sequence. To prevent such effect, the last sequence segment is treated as an additional batch, disregarding the number of its included frames. All experiments are preformed on a laptop platform with an Intel i7-3740QM CPU of 2.7GHz and a memory of 8GB. The proposed approach is named as JCSTD and implemented in C++ and runs in a single thread[3].

[2] Most parameter settings in the proposed approach are consistent with those approaches using a similar spatial or temporal constraint. Thus, a fair comparison is guaranteed.

[3] In the published work [145], the proposed approach is implemented in MATLAB. For application in automated driving system, it is refactored by C++ and the corresponding runtime measurement is also updated.

5.3.2 Ablation Study on Batch Process

Up to now, there are still two parameters remaining unsettled, i.e., the batch size f_l and f_t. However, the selection of both parameters is not trivial, because they both strongly affect the association accuracy. In the proposed approach, the batch size f_l corresponds to the maximal length of generated tracklet, with the assumption that the object velocity is nearly constant across included frames. Therefore, the parameter f_l should be well selected to not breach this assumption. A feasible idea is to determine the batch size f_l according to the frame rate of tested video. As sequences from evaluated datasets are recorded differently, the optimal batch size f_l should be searched based on experiments. Simultaneously, the optimal batch size f_t is searched in the same experiment due to its relationship with f_l (i.e., $f_t = N f_l$) revealed in Section 5.2.4. Here the training data from three benchmarks is utilized due to that groundtruth labels for their testing data are not available. For each dataset, the parameters f_l and N are respectively tested in a range of $5 \leq f_l \leq 30$ and $2 \leq N \leq 12$. The corresponding MOTA values in terms of both parameters are illustrated by heatmap in Fig. 5.3. From experimental results, it is clear that the optimum batch sizes for each dataset are strongly correlated with the frame rate, the camera status and the observed object behavior, discussed as follows:

KITTI: Although sequences are recorded with a unified frame rate of 10fps, since the camera moves over time, the motion model of constant velocities for objects in image is only valid in a few number of frames. As a result, the optimal MOTA value is situated at a very small batch size of $f_l = 5$.

MOT16: Most sequences in this dataset are recorded by frame rate range of 25 ~ 30fps. Since pedestrians are observed in very crowded scenarios, the motion estimation for long periods can be troublesome due to frequently occurred inter-class occlusion. Hence, a small batch size $f_l = 10$ performs better in this case (Fig. 5.3 (f)).

UA-DETRAC: Video sequences in this dataset are recorded with a unified frame rate of 25fps. As the camera observes the traffic flow, especially the vehicles on the road which are usually with a high density, the object number varies quickly over time. Hence, the optimum is located at a small batch size of $f_l = 10$.

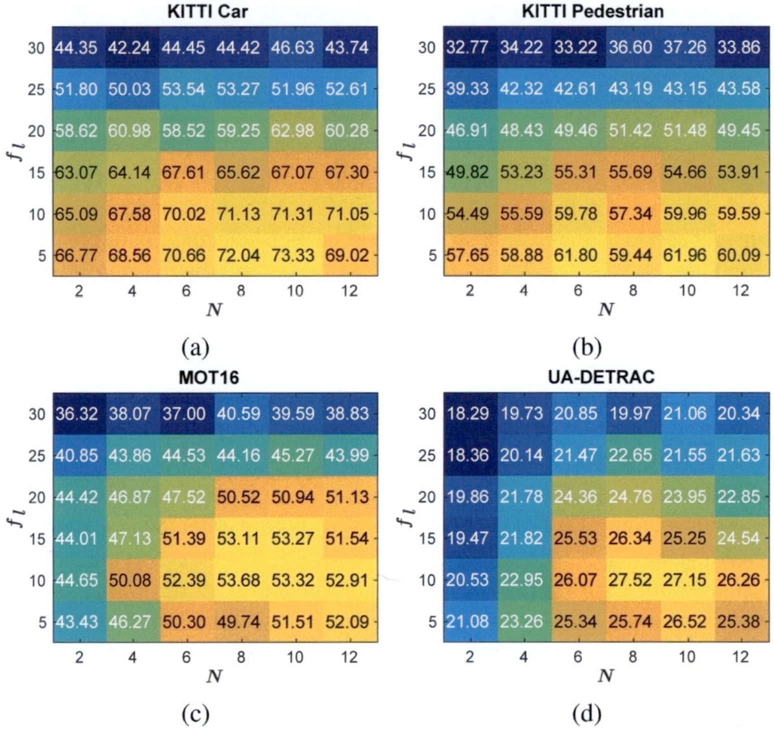

Figure 5.3: For each dataset, parameters f_l and N are respectively tested in a range of $5 \leq f_l \leq 30$ and $2 \leq N \leq 12$. Corresponding MOTA values in terms of both parameters are depicted in the form of heatmap.

Unlike the batch size f_l, the optimal parameter N of all datasets is mostly located in the range of 8 to 10. An explanation is that in tested sequences, the maximal duration of interrupted case, e.g., the full occlusion, is less than this length, which enables previously disappeared targets to be reidentified within such a big window. Hence, the optimal parameter N as well as the batch size f_l is kept for each dataset in further experiments.

5.3.3 Evaluation on Varied Dynamics

In automated driving systems, camera-based sensors are usually installed on a moving platform, i.e., the vehicle, and applied in both static and dynamic scenarios. Thus, this experiment is set to demonstrate the robustness of the proposed approach in scenarios with different camera dynamics. The related dataset is the KITTI benchmark. Since in this dataset, both pedestrians and vehicles are available, the performance of the proposed approach is also evaluated on different object classes. The results for each class are respectively reported in Table 5.2. As mentioned, in this dataset the camera is mounted on the roof of a moving vehicle and mainly observes its frontal areas. Hence, in the image, objects which are far away from the camera can be easily occluded by those in nearby regions. Moreover, the motion of a single object is also hard to estimate, because it is correlated with the motion of the vehicle. In spite of these challenging issues, from the results, it can be seen that the proposed method still surpasses the current top method MCMOT-CPD [94] by about 2% in terms of MOTA score for tracking cars. Among those compared methods, two of them are related to the proposed approach. The first one is the method SCEA [169]. It employs a similar spatial constraint which considers the local relationship between tracked targets. However, as temporal constraints are not integrated, it performs inferior to the proposed method especially in associating long ranged objects. The second one is the method SSP [95] which also batch-wisely processes tracks but within a flow network. Since it mainly considers association in consecutive frames, compared with proposed method, it has difficulties especially in dealing with occlusions. For tracking pedestrians, the method MDP [168] ranks on the top. It is a reinforcement learning method, which can better learn the complex behavior of targets such as pedestrians. Since optical flow or deep feature is exploited in NOMT [29] and MCMOT-CPD [94], they also achieve a higher accuracy than the proposed approach, which demonstrates that high-level features indeed benefit the tracking of non-rigid targets, e.g., pedestrians. Although in the proposed method, only hand-crafted feature like the HOG [49] is employed, it achieves a MOTA value with small gap to the top performing methods and ranks at the second place in terms of MOTP value. Additionally, the proposed method also achieves a very low number of ID switches as well as low ML scores, which further demonstrates the effectiveness of joint constraints.

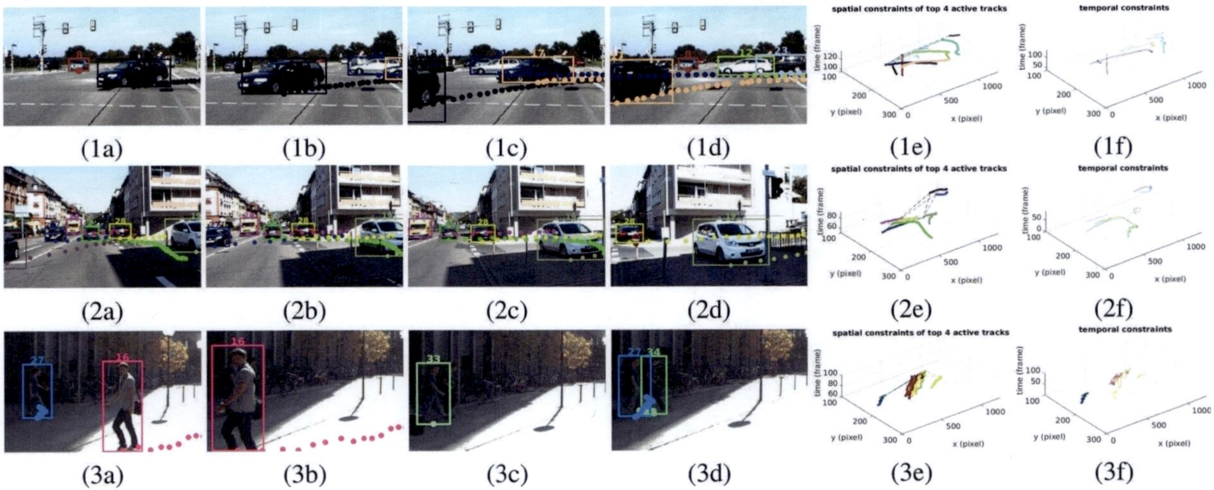

Figure 5.4: Examples of tracking vehicles and pedestrians with dynamic and static cameras. The sequences are from the KITTI benchmark. The first four columns show the tracked objects, which are denoted in different colors along with their IDs. The target in current frame is indicated by a bounding box and its trajectory in last 30 frames are represented by points. In the last two columns spatial and temporal constraints are respectively displayed with tracks in a 3-D space with the z-axis indicating the frame numbers and the x,y-axes denoting the image coordinates. Spatial constraints here are represented by pair-wise motion patterns. The selected track states are pair-wise linked by dash lines and shown only for the top 4 active tracks. For temporal constraints, the association between tracks is indicated by a black line while tracks without associations are denoted in transparent colors.

Data	Method	Setting	MOTA ↑	MOTP ↑	MT ↑	ML ↓	IDS ↓	FR ↓	Time ↓	Hardware
KITTI Car	MCMOT-CPD [94]		78.90%	**82.13%**	52.31%	11.69%	228	536	**0.01s**	1×3.5GHz
	NOMT [29]		78.15%	79.46%	**57.23%**	13.23%	**31**	**207**	0.09s	16×2.5GHz
	LP-SSVM [156]		77.63%	77.80%	56.31%	8.46%	62	539	0.02s	1×2.5GHz
	MDP [168]	online	76.59%	82.10%	52.15%	13.38%	130	387	0.9s	8×3.5GHz
	SCEA [169]	online	75.58%	79.39%	53.08%	11.54%	104	448	0.06s	1×4GHz
	CIWT [120]	online	75.39%	79.25%	49.85%	10.31%	165	660	0.28s	1×2.5GHz
	SSP [95]		72.72%	78.55%	53.85%	8.00%	185	932	0.6s	1×2.7GHz
	DCO-X [112]		68.11%	78.85%	37.54%	14.15%	318	959	0.9s	1×3.5GHz
	RMOT [170]	online	65.83%	75.42%	40.15%	9.69%	209	727	0.02s	1×3.5GHz
	ODAMOT [54]	online	59.23%	75.45%	27.08%	15.54%	389	1274	1s	1×2.5GHz
	TBD [171]		55.07%	78.35%	20.46%	32.62%	**31**	529	10s	1×2.5GHz
	CEM [111]		51.94%	77.11%	20.00%	31.54%	125	396	0.09s	1×3.5GHz
	JCSTD	online	**80.57%**	81.81%	56.77%	**7.38%**	61	643	0.07s	1×2.7GHz
KITTI Pedestrian	MDP [168]	online	**47.22%**	70.36%	24.05%	**27.84%**	87	825	0.9s	8×3.5GHz
	NOMT [29]		46.62%	71.45%	**26.12%**	34.02%	63	666	0.09s	16×2.5GHz
	MCMOT-CPD [94]		45.94%	**72.44%**	20.62%	34.36%	143	764	**0.01 s**	1×3.5GHz
	SCEA [169]	online	43.91%	71.86%	16.15%	43.30%	56	641	0.06s	1×4.0GHz
	RMOT [170]	online	43.77%	71.02%	19.59%	41.24%	153	748	0.02s	1×3.5GHz
	LP-SSVM [156]		43.76%	70.48%	20.62%	34.36%	73	809	0.02s	1×2.5GHz
	CIWT [120]	online	43.37%	71.44%	13.75%	34.71%	112	901	0.28s	1×2.5GHz
	CEM [111]		27.54%	68.48%	8.93%	51.89%	96	**608**	0.09s	1×3.5GHz
	JCSTD	online	44.20%	72.09%	16.49%	33.68%	**53**	917	0.07s	1×2.7GHz

Table 5.2: Evaluation of approaches on tracking cars and pedestrians on the KITTI benchmark till April 2018. The up-arrow means that higher values are better while a down-arrow implies that lower values are preferred. The proposed method and best values are marked in bold.

Some tracking examples are presented in Fig. 5.4. In the first sequence, the vehicle with ID 8 (brown) is displayed in frame (1a) while a few cars are moving across the intersection and occlude it in frame (1b) and (1c). Despite multiple occlusions, the vehicle 8 is successfully reidentified in frame (1d) based on the analysis of motion pattern among objects, illustrated in (1e). In the second sequence, the camera along with the ego-vehicle sharply turns right. Consequently, observed objects in the image move towards to the left boundary (frame (2a)-(2d)). Although object positions change quickly in the image, since the association is based on pair-wise motion patterns, no target is lost. In the third sequence, a slowly walking pedestrian with the ID 27 (blue) is detected in frame (3a), which can be regarded as a nearly static object. As the vehicle moves forward, pedestrian 27 is occluded by another with ID 16 (pink).

In frame (3c), the occlusion disappears but the pedestrian 27 is regarded as a new object with the ID 33 (green). By performing constrained temporal association, depicted in (3f), it is remapped to object 27 in frame (3d).

5.3.4 Evaluation on Varied Platforms

In the next experiment, the proposed approach is tested on the MOT16 benchmark to validate its performance on different platforms. In this dataset, video sequences are recorded in a big range of view angles: from a small moving trolley to elevated position on big transportation tools like trams. The main observed objects are pedestrians walking in crowds. Compared with the previous experiment, the object density in these images are much higher, e.g., up to 70 pedestrians can appear in one frame. To release the computational burden and maintain a fast processing speed, the maximal number of stored pair-wise motion patterns for each object is limited to 10 in this experiment. A random sampling is also conducted to choose these motion patterns. Tracking results are reported in Table 5.3. According to the results, it can be seen that the proposed approach achieves an acceptable MOTA score, which is slightly less than that of the top method FWT [72] and comparable to the second one NLLMPa [96]. In terms of online methods, the result of proposed method is more satisfying, which ranks on the top comparing with other state-of-the-art approaches. Moreover, the proposed approach achieves the best score on the ratio of mostly lost targets (ML) among all compared approaches, which implies that most of the targets are identified by the proposed approach. Another phenomenon can be seen from the table is that the average number of ID switches of online methods (including the proposed approach) is higher than those offline methods such as FWT [72] and NLLMPa [96]. This is due to the fact that offline methods always consider detection hypotheses over the entire sequence while online methods only deal with detections up to current frame. As the observation window is greatly limited in online methods, the average track length turns out to be much shorter, which can be inferred from those MT scores and fragment numbers. Hence, the association for tracks becomes suboptimal without access to global information. Another fact is that the proposed method only utilizes a small number of motion patterns to construct spatial constraints, which also makes the association less robust, particularly for non-rigid objects like pedestrians. Nevertheless, considering that the MOTA gap between the proposed approach and the top method FWT is only 0.4%, the minor sacrifice in accuracy leads to significant boost in run time performance, i.e., about 10fps, which is suitable for most real time applications.

Data	Method	Setting	MOTA ↑	MOTP ↑	MT ↑	ML ↓	IDS ↓	FR ↓	Time ↓	Hardware
MOT 16	FWT [72]		**47.8%**	75.5%	19.1%	38.2%	852	1534	1.67s	8×3.5GHz
	NLLMPa [96]		47.6%	**78.5%**	17.0%	40.4%	629	768	0.12s	1×3.6GHz
	MDPNN16 [131]	online	47.2%	75.8%	14.0%	41.6%	774	1675	1s	1×3GHz
	MCjoint [81]		47.1%	76.3%	**20.4%**	46.9%	370	598	1.67s	1×2.4GHz
	NOMT [29]		46.4%	76.6%	18.3%	41.4%	**359**	**504**	0.38s	16×2.4GHz
	JMC [142]		46.3%	75.7%	15.5%	39.7%	657	1114	1.25s	1×3.2GHz
	MHT_DAM [85]		45.8%	76.3%	16.2%	43.2%	590	781	1.25s	12×3.6GHz
	CDA_DDALv2 [9]	online	43.9%	74.7%	10.7%	44.4%	676	1795	2s	1×3.1GHz
	oICF [84]	online	43.2%	74.3%	11.3%	48.5%	381	1404	2.5s	2×2.3GHz
	LNF1 [46]		41.0%	74.8%	11.6%	51.3%	430	963	0.24s	8×2.7GHz
	EAMTT_pub [133]	online	38.8%	75.1%	7.9%	49.1%	965	1657	**0.08s**	1×3.4GHz
	OVBT [11]	online	38.4%	75.4%	7.5%	47.3%	1321	2140	0.33s	1×2.5GHz
	JCSTD	online	47.4%	74.4%	14.4%	**36.4%**	1266	2696	0.11 s	1×2.7GHz

Table 5.3: Evaluation of approaches for tracking pedestrians on the MOT16 benchmark till April 2018. The up-arrow means that higher values are better while a down-arrow implies that lower values are preferred. The proposed method and best values are marked in bold.

Tracking examples are displayed in the first two sequences of Fig. 5.5. The first video is captured by a camera installed on a small trolley. Two pedestrians respectively with the ID 4 (pink) and 5 (yellow) are crossing each other in the image. Despite that pedestrian 4 is occluded in frame (1b), leveraging the spatially constrained motion pattern, it is reidentified in frame (1c) and (1d). In the second sequence, the camera is installed on a tram in an elevated position. The tram sharply turns left while objects in the image are moving to the opposite direction (frame (2a)-(2d)). Despite that, targets are still successfully tracked, e.g., pedestrian 99 (purple) can be recovered after the occlusion. As such association can be directly solved by the spatial constraint, the temporal constraint here is not activated.

5.3.5 Evaluation on Synthesized Approaches

In the previous chapters, we already introduced trackers that can handle severe occlusion and deteriorated vision conditions. However, they mainly focus on tracking of a single target. In this part, we explore their performance for multi-object tracking by integrating them in the proposed approach of this chapter. The utilized dataset is the UA-DETRAC benchmark [159], in which the camera is installed at a very high position to monitor the traffic flow. Since images are recorded under varying illumination and weather conditions, they are suitable to test the synthesized tracking approach. Here, we prepare two versions of synthesized approach, respectively denoted as JSCTD_O and JCSTD_V. The first version is integrated with the tracker for handling severe occlusion while the second one is integrated with the tracker for handling deteriorated vision[4]. In both versions, we initialize a separate tracker for each target and replace the matching operation (5.3) in the naive JCSTD with the correlation filter. The tracker is reinitialized when a new detection hypothesis is associated with the target. Otherwise, the state of the target is estimated by the tracker. The tracking results are reported in Table 5.4.

Data	Method	MOTA ↑	MOTP ↑	MT ↑	ML ↓	IDS ↓	FR ↓	fps ↑	Hardware
UA-DETRAC	GOG [124]	14.2%	37.0%	13.9%	19.9%	3334.6	3172.4	**389.51**	4×2.9GHz
	CMOT [8]	12.6%	36.1%	16.1%	18.6%	285.3	1516.8	3.79	4×2.9GHz
	H^{2}T [160]	12.4%	35.7%	14.8%	19.4%	852.2	1117.2	3.02	4×2.9GHz
	IHTLS [42]	11.1%	36.8%	13.8%	19.9%	953.6	3556.9	19.79	4×2.9GHz
	DCT [4]	10.8%	37.1%	6.7%	29.3%	**141.4**	**132.4**	2.19	4×2.9GHz
	CEM [3]	5.1%	35.2%	3.0%	35.3%	267.9	352.3	4.62	4×2.9GHz
	JCSTD	17.0%	36.9%	16.6%	17.6%	480.2	1611.5	44.09	1×2.7GHz
	JCSTD_O	17.3%	36.8%	16.8%	17.4%	478.4	1603.7	10.25	1×2.7GHz
	JCSTD_V	**17.8%**	**37.2%**	**17.5%**	**17.0%**	473.6	1594.2	13.14	1×2.7GHz

Table 5.4: Evaluation of tracking approaches on the UA-DETRAC till April 2018. To be consistent with the original benchmark, the run time performance is given in fps. The up-arrow means that higher values are preferred while a down-arrow implies that lower values are better. The proposed methods and best values are marked in bold.

From the results, it is clear that the average precision of all tracking approaches is apparently lower than in the other two datasets. This can be explained by the

[4] At the moment, the trackers for handling severe occlusion and deteriorated vision cannot be integrated together. Because the first one employs a segmentation approach, the recognized object/part is in a pixel-level precision, which can result in historical samples with greatly varied shapes and sizes, thus making temporal optimization of the second tracker difficult.

relatively poor quality of detector utilized in UA-DETRAC. Since the camera is installed on a high position, the image is captured nearly from a top view. Hence, detection approaches, mainly learned for frontal views, can only yield a suboptimal performance. Moreover, vehicles with deformable shapes such as buses or trucks with multiple sections or containers frequently appear in the image but are difficult to be precisely detected. In spite of these factors, the proposed approach JCSTD still outperforms other state-of-the-art methods in terms of most metrics. By integrating the tracker for deteriorated vision, the JCSTD_V achieves the top performance in four metrics, i.e., MOTA, MOTP, MT and ML. Especially in the first and fourth metric, the gain is about 1%, which is benefited from a better object matching in deteriorated vision (with examples shown in Fig. 5.5 (3) and (4)). In comparison, by integrating the tracker with severe occlusion handling in JCSTD_O, the performance is not much boosted as in JCSTD_V. The reason is that in these nearly top view images, the occlusion ratio is not so high as in previous experiments and some partially occluded objects can be directly recognized by the reference detector (see Fig. 5.5 (3a)-(3d)). Thus, the profit by employing the occlusion tracker is limited. On other metrics, the DCT method [4] shows the lowest number in both ID switches and fragments. However, this result is attained at the cost that only a few objects are persistently tracked (low MT ratio) while a plenty of true targets are lost (high ML ratio). In contrast, by incorporating the joint constraint in JCSTD as well as its extended versions, the ratio of tracked targets is significantly increased while the number of lost objects is little. For the runtime performance, it can be seen that by integrating external trackers, the speed of naive JCSTD declines about 30fps in JCSTD_O and JCSTD_V. However, their speed are still comparable to the naive JCSTD in previous experiments, which is due to the fact that the object number in UA-DETRAC is smaller than in the other two datasets. The GOG approach [124] is the fastest one with a speed of less than 400fps. This approach utilizes a flow network and searches the min-cost flow by a fast implemented dynamic programming. As it only considers the motion of a single object and conducts association mainly between successive frames, more association errors occur, e.g., in ID switch, than the joint constraint-based method, which further proves the effectiveness of proposed approach.

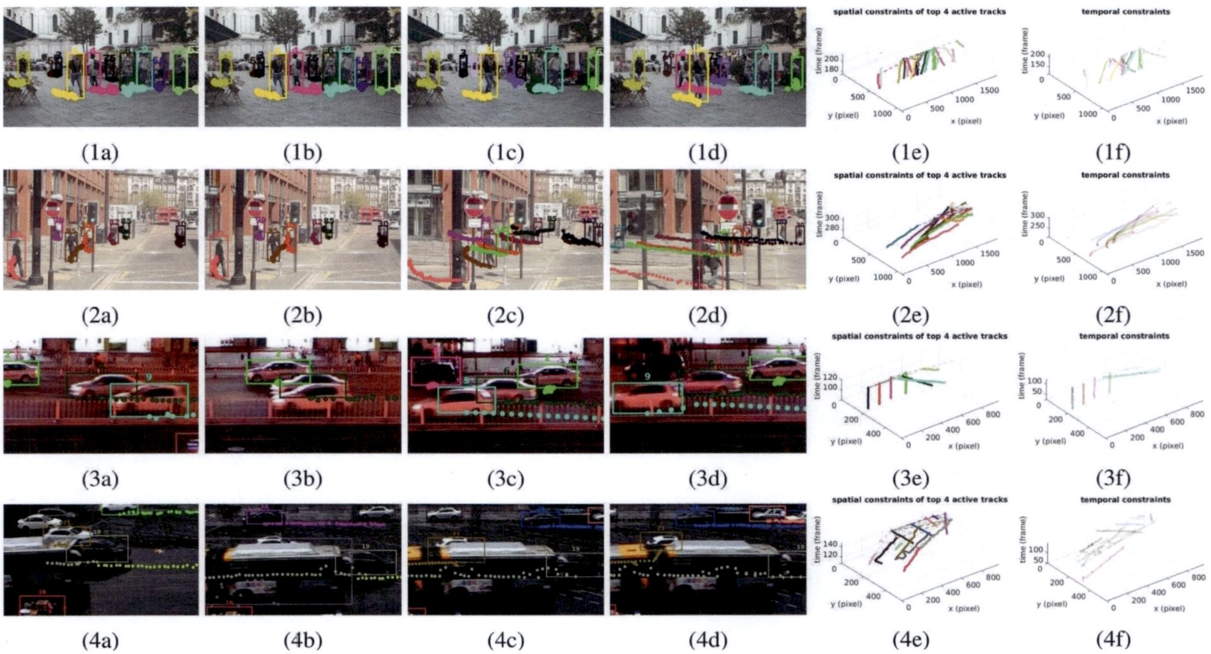

Figure 5.5: Examples of tracking with varied platforms and weather factors displayed in the same layout as in Fig. 5.4. The first two sequences are from the MOT16 benchmark while the last two are from UA-DETRAC.

5.3.6 Runtime Performance Analysis

To reveal crucial impact factors on the runtime performance, the average time cost over all datasets is plotted for each stage of the proposed approach as well as for the total procedure in terms of object number in Fig. 5.6. From the second plot, it can be seen that in the proposed approach the time cost for tracklet creation is linear to the object number. This is due to the fact that the tracklet linearly grows with its assigned detection hypotheses. As the maximal time never exceeds 4 milliseconds, it is negligible in comparison with other stages. In contrast, the association with spatial constraints is the most time-dominant. As motion patterns are created pair-wisely, the curve tends to fit an exponential function along with increased object number (see Fig. 5.6 (c)). As for the temporal constraint, its processing time is much shorter and almost equals the half of the time by the spatial constraint, as shown in Fig. 5.6 (d). This can be credited to the AL-ICM approach, which is capable to solve large association problem at a relative low computational burden. Regarding the total processing time, a speed of nearly 10 fps can be achieved if no more than 50 objects are tracked. Although our approach is not the fastest one in all experiments, it is still capable to run in real time. Note that the proposed method runs only in a single core, there is still space for improvement. Considering utilizing multi-threads or GPU implementation, a faster processing speed is still possible to be achieved.

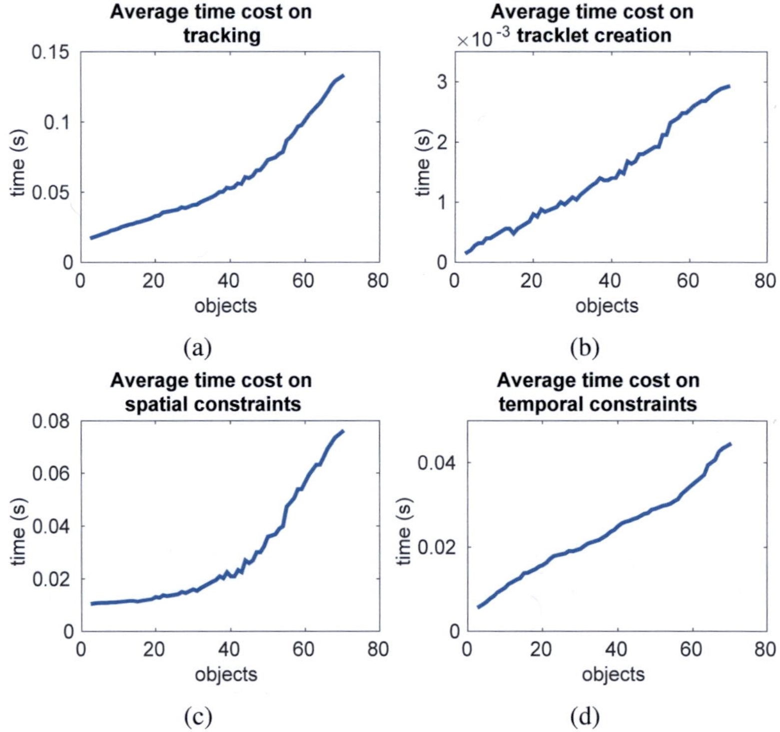

Figure 5.6: Plot (a) shows the average time cost of the proposed tracking approach over all datasets in terms of object number. Plots (b) to (d) respectively show the average time cost for each individual stage: tracklet creation, association by spatial and temporal constraint.

6 Conclusion and Outlook

This thesis has addressed challenging problems of visual object tracking in traffic scenarios, i.e., severe occlusion, deteriorated vision and multi-object reidentification. For solving these problems, novel tracking solutions which aggregate the information in various levels from image feature to object parts/groups are proposed in this thesis. All these solutions are only based on image sequence captured by a monocular camera and do not require additional sensors.

To track objects undergoing severe occlusion, a novel tracker employing part filters is presented in Chapter 3. By analyzing the variation of appearance model and filter response, this tracking approach can successfully recognize the occurrence of occlusion. Unlike existing approaches, in this tracker, both the number and size of part filters are determined quite flexibly and thus adapted to visible areas of the target. Relying on a masking process, this tracker can provide a pixel-level precision in distinguishing the fore- and background image regions. A color prior is also embedded into the final response map to boost the inference of full object tracker. Extensive experiments have demonstrated that in comparison with state-of-the-art approaches, the proposed tracker has achieved great success in dealing with severe occlusion, particularly the long-termed ones. A further study on more generalized scenarios revealed that the proposed tracker performs outstandingly in tracking various object classes under varied conditions, which has verified the generalization ability of the proposed tracker. Since this tracker and its utilized part filters are based on the technique of the kernelized correlation filter, the vast amount of computation can be completed by element-wise operation in the frequency domain. In cooperation with a dynamic filter management framework, the tracker provides a fast processing speed for real time applications.

In Chapter 4, a novel tracker employing both channel-wise and temporal weighting is presented for object tracking with deteriorated vision such as caused by low environmental illumination or adverse weather. The tracker is tailored from the baseline framework of the KCF but decomposes visual features into several small expert filters. Through exploring the reliability of each expert, the proposed tracker is able to extract the most discriminative visual features. By searching confidential training samples in the time domain and integrating the human memory model, which is interpreted by the forgetting curve, the tracker can be learned without the influence of corrupted samples caused by contaminated vision. Both above steps are successfully incorporated in one learning framework. Due to the biconvexity property of the objective function, both problems can be jointly solved by the Alternate Convex Search method. Experiments on image datasets captured under low illumination conditions demonstrate the improved state-of-the-art performance of the proposed tracker. A more comprehensive study reveals that the proposed tracker performs robustly against various challenging factors such as color, weather, time, vehicle class and behavior. Leveraging an elaborate design, the proposed tracker can be directly learned in the frequency domain, only with little computational burden. Thus, this tracker also permits a real time performance.

The first two approaches mainly focus on tasks related to tracking of a specific target, e.g., estimating the trajectory of a partially occluded pedestrian or following the car ahead of the ego-vehicle during night. In other traffic scenarios, it also requires to analyze the behavior of multiple objects based on their trajectories, e.g., for surveillance purpose. Such task is related to multi-object tracking, which can be cast to the object reidentification/association problem, thanks to recent progress in object detection techniques. However, existing methods still have difficulties in dealing with camera motion and ambiguities between long ranged objects. Regarding these issues, a novel multi-object tracking approach within a unified framework to solve both challenges is presented in Chapter 5. This approach is based on the strategy of joining both the spatial and temporal constraints. The first constraint analyzes pair-wise motion pattern between objects to predict the target location while the second one relies on a subgraph-based model to recover long time vanished objects. To maintain a seamless cooperation between both constraints, a 3-stage scheme within an alternative optimization fashion is proposed for computational efficiency. Additionally, new rules are introduced to effectively deal with association for tracklets-to-tracks, so that the processing time is decoupled from the length of tracklet. By extensive experiments with setups in varied camera dynamics and view angles, the proposed approach has been verified with comparable or even improved state-of-the-art performance. This approach is also successfully integrated with trackers presented in previous chapters and exhibits a promoted tracking precision in more challenging test scenarios. Considering the run time performance, the proposed approach is able to run online with a speed of more than 10fps in most cases.

In the future research, deep networks can be utilized to learn a better appearance representation of tracked targets, especially in the case of occlusion recognition, as deep features can encode much richer information than conventional hand-crafted ones. Moreover, for a more accurate identification of visible object parts, convolutional neural networks can also be adopted in the masking process, since they have achieved more encouraging results for recent semantic segmentation tasks. For handling deteriorated vision, the presented tracking approach in this thesis mainly focuses on the algorithmic level, which is decoupled from the imaging process. Thus, approaches for enhancing the image quality can be integrated into the proposed tracker to boost the feature selection as well as learning the classifier. Employing additional sensors such as the lidar can also provide valuable information to help the target perception especially in low illuminated scenarios. Normally, surveillance tasks are more interested in the locations or trajectories of targets in the real world. Hence, an extension of current multi-object tracking framework is to integrate the 3-D environmental information based on depth sensors such as the high precision radar or lidar. Since an automated driving vehicle is usually mounted with multiple cameras to perceive its surroundings in a full angle, object tracking across images by different cameras is a promising direction for further development of the proposed tracking framework. Nevertheless, the computation amount increases along with the complexity of the exploited approach, which in turn negatively impacts its runtime performance. Thanks to the rapid development of processor industries, this problem can still be solved by deploying advanced hardware, such as high-end GPUs.

A AL-ICM Algorithm

The complete AL-ICM algorithm utilized in this thesis is presented as follows:

Algorithm 1 The complete AL-ICM algorithm.

Input: Square cost matrix $\boldsymbol{\Omega}$ of size $n \times n$, iteration number N_{iter};
Output: Label vector $\boldsymbol{l}_{out}$;

Procedure:
Set initial labels $\boldsymbol{l} = \text{ones}(n, 1)$ and initial iteration $i = 1$;
For each iteration $i \leq N_{iter}$, do
    Generate new labels $\boldsymbol{l}_{new} = \text{ICM}(\boldsymbol{\Omega}, \boldsymbol{l})$;
    Adjust $\boldsymbol{l}_{new}$ so that its smallest label is equal to one and
    all its labels are successive nature numbers.
    If $\boldsymbol{l}_{new}$ equals $\boldsymbol{l}$
        break;
    End
    $\boldsymbol{l} = \boldsymbol{l}_{new}$;
End
$\boldsymbol{l}_{out} = \boldsymbol{l}$;

Algorithm 2 The ICM algorithm.

Input: Square cost matrix $\boldsymbol{\Omega}$ of size $n \times n$, label vector $\boldsymbol{l}$ with a total number of n;
Output: Label vector $\boldsymbol{l}_{new}$;

Procedure:
Initialize new label vector $\boldsymbol{l}_{new}$ = ones$(n, 1)$;
Get maximum label l_{max} = max$(\boldsymbol{l})$;
For each column $i = 1 : n$, do
  Initial buffer $\boldsymbol{b}$ = zeros$(l_{max}, 1)$;
  For each row $j = 1 : n$, do
    If $j = i$
      continue;
    End
    $l_{tmp} = \boldsymbol{l}_{new}(j)$ and $v = \boldsymbol{\Omega}(j, i)$;
    If $v < 0$
      $\boldsymbol{b}(l_{tmp}) += v$;
    End
  End
  Set $m_i = l_{max} + 1$ and $m_x = 0$;
  For each $k = 1 : l_{max}$, do
    If $\boldsymbol{b}(k) < m_x$
      $m_x = \boldsymbol{b}(k)$;
      $m_i = k$;
    End
  End
  $\boldsymbol{l}_{new}(i) = m_i$;
  If $m_i > l_{max}$
    $l_{max} = m_i$;
  End
End

Bibliography

[1] O. Akin, E. Erdem, A. Erdem, and K. Mikolajczyk, "Deformable part-based tracking by coupled global and local correlation filters," *Journal of Visual Communication and Image Representation*, vol. 38, pp. 763–774, 2016, Elsevier, Amsterdam.

[2] D. Alspach, "A gaussian sum approach to the multi-target identification-tracking problem," *Automatica*, vol. 11, no. 3, pp. 285–296, 1975, Elsevier, Amsterdam.

[3] A. Andriyenko and K. Schindler, "Multi-target tracking by continuous energy minimization," in *IEEE Conference on Computer Vision and Pattern Recognition (CVPR)*, June 2011, pp. 1265–1272, IEEE, Colorado Springs, DOI:10.1109/CVPR.2011.5995311.

[4] A. Andriyenko, K. Schindler, and S. Roth, "Discrete-continuous optimization for multi-target tracking," in *IEEE Conference on Computer Vision and Pattern Recognition (CVPR)*, June 2012, pp. 1926–1933, IEEE, Providence, DOI:10.1109/CVPR.2012.6247893.

[5] R. C. Atkinson and R. M. Shiffrin, "Human memory: A proposed system and its control processes," *Psychology of learning and motivation*, vol. 2, pp. 89–195, 1968, Elsevier, Amsterdam.

[6] S. Avidan, "Ensemble tracking," *IEEE Transaction on Pattern Analysis and Machine Intelligence (PAMI)*, vol. 29, no. 2, pp. 261–271, 2007, IEEE, DOI:10.1109/TPAMI.2007.35.

[7] B. Babenko, M.-H. Yang, and S. Belongie, "Visual tracking with online multiple instance learning," in *IEEE Conference on Computer Vision and Pattern Recognition (CVPR)*, 2009, pp. 983–990, IEEE, Miami, DOI:10.1109/CVPR.2009.5206737.

[8] S.-H. Bae and K.-J. Yoon, "Robust online multi-object tracking based on tracklet confidence and online discriminative appearance learning," in *IEEE conference on Computer Vision and Pattern Recognition (CVPR)*, June 2014, pp. 1218–1225, IEEE, Columbus, DOI:10.1109/CVPR.2014.159.

[9] S.-H. Bae and K.-J. Yoon, "Confidence-based data association and discriminative deep appearance learning for robust online multi-object tracking," *IEEE Transactions on Pattern Analysis and Machine Intelligence (PAMI)*, vol. 40, no. 3, pp. 595–610, 2017, IEEE, DOI:10.1109/TPAMI.2017.2691769.

[10] S. Bagon and M. Galun, "Large scale correlation clustering optimization," in *CoRR*, vol. abs/1112.2903, 2011, arXiv. http://arxiv.org/abs/1112.2903.

[11] Y. Ban, S. Ba, X. Alameda-Pineda, and R. Horaud, "Tracking Multiple Persons Based on a Variational Bayesian Model," in *European Conference on Computer Vision Workshops (ECCVW)*, 2016, pp. 52–67, Springer, Berlin, DOI:10.1007/978-3-319-48881-3_5.

[12] N. Bansal, A. Blum, and S. Chawla, "Correlation clustering," *Machine Learning*, vol. 56, no. 1-3, pp. 89–113, 2004, Kluwer Academic Publishers, Dordrecht.

[13] J. L. Barron, D. J. Fleet, and S. S. Beauchemin, "Performance of optical flow techniques," in *IEEE Conference on Computer Vision and Pattern Recognition (CVPR)*, 1994, pp. 43–77, IEEE, Champaign, DOI:10.1109/CVPR.1992.223269.

[14] BAST, "Traffic and Accident Data Summary Statistics - Germany 2017," Oct. 2017, Federal Highway Research Institute, Bergisch Gladbach.

[15] H. Bay, T. Tuytelaars, and L. Van Gool, "Surf: Speeded up robust features," in *European Conference on Computer Vision (ECCV)*, 2006, pp. 404–417, Springer, Berlin, DOI:10.1007/11744023_3.

[16] J. Berclaz, F. Fleuret, E. Turetken, and P. Fua, "Multiple object tracking using k-shortest paths optimization," *IEEE Transaction on Pattern Analysis and Machine Intelligence (PAMI)*, vol. 33, no. 9, pp. 1806–1819, 2011, IEEE, DOI:10.1109/TPAMI.2011.21.

[17] M. J. Black and A. D. Jepson, "Eigentracking: Robust matching and tracking of articulated objects using a view-based representation," *International Journal of Computer Vision (IJCV)*, vol. 26, no. 1, pp. 63–84, 1998, Springer, Berlin, DOI:10.1023/A:1007939232436.

[18] D. S. Bolme, J. R. Beveridge, B. A. Draper, and Y. M. Lui, "Visual object tracking using adaptive correlation filters," in *IEEE Conference on Computer Vision and Pattern Recognition (CVPR)*, 2010, pp. 2544–2550, IEEE, San Francisco, DOI:10.1109/CVPR.2010.5539960.

[19] M. D. Breitenstein, F. Reichlin, B. Leibe, E. Koller-Meier, and L. V. Gool, “Online multiperson tracking-by-detection from a single, uncalibrated camera,” *IEEE Transactions on Pattern Analysis and Machine Intelligence (PAMI)*, vol. 33, no. 9, pp. 1820–1833, Sept 2011, IEEE, DOI:10.1109/TPAMI.2010.232.

[20] J. V. Brummelen, M. O’Brien, D. Gruyer, and H. Najjaran, “Autonomous vehicle perception: The technology of today and tomorrow,” *Transportation Research Part C: Emerging Technologies*, vol. 89, pp. 384 – 406, 2018, Elsevier, Amsterdam.

[21] Z. Cai, Q. Fan, R. S. Feris, and N. Vasconcelos, “A unified multi-scale deep convolutional neural network for fast object detection,” in *European Conference on Computer Vision (ECCV)*, 2016, pp. 354–370, Springer, Cham, DOI:10.1007/978-3-319-46493-0_22.

[22] M. Calonder, V. Lepetit, C. Strecha, and P. Fua, “Brief: Binary robust independent elementary features,” in *European Conference on Computer Vision (ECCV)*, 2010, pp. 778–792, Springer, Berlin, DOI:10.1007/978-3-642-15561-1_56.

[23] D. C. Champeney, “A handbook of fourier theorems,” 1987, Cambridge University Press, Cambridge.

[24] C. Chen, Q. Chen, J. Xu, and V. Koltun, “Learning to see in the dark,” in *CoRR*, vol. abs/1805.01934, 2018, arXiv. http://arxiv.org/abs/1805.01934.

[25] L. Chen, X. Hu, T. Xu, H. Kuang, and Q. Li, “Turn signal detection during nighttime by cnn detector and perceptual hashing tracking,” *IEEE Transactions on Intelligent Transportation Systems*, vol. 18, no. 12, pp. 3303–3314, 2017, IEEE, DOI:10.1109/TITS.2017.2683641.

[26] Y.-L. Chen, Y.-H. Chen, C.-J. Chen, and B.-F. Wu, “Nighttime vehicle detection for driver assistance and autonomous vehicles,” in *IEEE International Conference on Pattern Recognition (ICPR)*, vol. 1, 2006, pp. 687–690, IEEE, Hongkong, DOI:10.1109/ICPR.2006.858.

[27] Z. Chen and A. Storjohann, “A blas based c library for exact linear algebra on integer matrices,” in *International Symposium on Symbolic and Algebraic Computation*, 2005, pp. 92–99, ACM, New York, DOI:10.1145/1073884.1073899.

[28] P. Chockalingam, N. Pradeep, and S. Birchfield, "Adaptive fragments-based tracking of non-rigid objects using level sets," in *IEEE International Conference on Computer Vision (ICCV)*, 2009, pp. 1530–1537, IEEE, Kyoto, DOI:10.1109/ICCV.2009.5459276.

[29] W. Choi, "Near-online multi-target tracking with aggregated local flow descriptor," in *IEEE International Conference on Computer Vision (ICCV)*, 2015, pp. 3029–3037, IEEE, Santiago, DOI:10.1109/ICCV.2015.347.

[30] N. Christianini and J. Shawe-Taylor, "An introduction to support vector machines and other kernel-based learning methods," *Robotica*, vol. 18, no. 6, pp. 687–689, 2000, Cambridge university press, Cambridge.

[31] D. Comaniciu, V. Ramesh, and P. Meer, "Real-time tracking of non-rigid objects using mean shift," in *IEEE Conference on Computer Vision and Pattern Recognition (CVPR)*, 2000, pp. 142–149, IEEE, Hilton Head Island, DOI:10.1109/CVPR.2000.854761.

[32] D. Comaniciu, V. Ramesh, and P. Meer, "Kernel-based object tracking," *IEEE Transactions on pattern analysis and machine intelligence (PAMI)*, vol. 25, no. 5, pp. 564–577, 2003, IEEE, DOI:10.1109/TPAMI.2003.1195991.

[33] G. Cookson and B. Pishue, "Inrix global traffic scorecard 2016," Feb. 2017, INRIX Reserach, Kirkland.

[34] J. W. Cooley, P. A. Lewis, and P. D. Welch, "The fast fourier transform and its applications," *IEEE Transactions on Education*, vol. 12, no. 1, pp. 27–34, 1969, IEEE, DOI:10.1109/TE.1969.4320436.

[35] N. Dalal and B. Triggs, "Histograms of oriented gradients for human detection," in *IEEE Conference on Computer Vision and Pattern Recognition (CVPR)*, June 2005, pp. 886–893, IEEE, San Diego, DOI:10.1109/CVPR.2005.177.

[36] M. Danelljan, G. Häger, F. Khan, and M. Felsberg, "Accurate scale estimation for robust visual tracking," in *British Machine Vision Conference (BMVC)*, Sept 2014, pp. 1–11, BMVA Press, Durham.

[37] M. Danelljan, G. Häger, F. S. Khan, and M. Felsberg, "Adaptive decontamination of the training set: A unified formulation for discriminative visual tracking," in *IEEE Conference on Computer Vision and Pattern Recognition (CVPR)*, 2016, pp. 1430–1438, IEEE, Las Vegas, DOI:10.1109/CVPR.2016.159.

[38] M. Danelljan, G. Hager, F. Shahbaz Khan, and M. Felsberg, "Learning spatially regularized correlation filters for visual tracking," in *IEEE International Conference on Computer Vision (ICCV)*, 2015, pp. 4310–4318, IEEE, Santiago, DOI:10.1109/ICCV.2015.490.

[39] M. Danelljan, A. Robinson, F. S. Khan, and M. Felsberg, "Beyond correlation filters: Learning continuous convolution operators for visual tracking," in *European Conference on Computer Vision (ECCV)*, 2016, pp. 472–488, Springer, Cham, DOI:10.1007/978-3-319-46454-1_29.

[40] M. Danelljan, F. Shahbaz Khan, M. Felsberg, and J. Van de Weijer, "Adaptive color attributes for real-time visual tracking," in *IEEE Conference on Computer Vision and Pattern Recognition (CVPR)*, 2014, pp. 1090–1097, IEEE, Columbus, DOI:10.1109/CVPR.2014.143.

[41] P. J. Davis, "Circulant matrices," 2012, American Mathematical Soc., Washington.

[42] C. Dicle, O. I. Camps, and M. Sznaier, "The way they move: Tracking multiple targets with similar appearance," in *IEEE International Conference on Computer Vision (ICCV)*, 2013, pp. 2304–2311, IEEE, Sydney, DOI:10.1109/ICCV.2013.286.

[43] T. B. Dinh, N. Vo, and G. Medioni, "Context tracker: Exploring supporters and distracters in unconstrained environments," in *IEEE Conference on Computer Vision and Pattern Recognition (CVPR)*, 2011, pp. 1177–1184, IEEE, Colorado Springs, DOI:10.1109/CVPR.2011.5995733.

[44] A. Dosovitskiy, J. T. Springenberg, and T. Brox, "Unsupervised feature learning by augmenting single images," in *CoRR*, vol. abs/1312.5242, 2013, arXiv. http://arxiv.org/abs/1312.5242.

[45] H. Ebbinghaus, "Memory: a contribution to experimental psychology," *Annals of neurosciences*, vol. 20, no. 4, pp. 155–156, 2013, Karger Publishers, Basel.

[46] L. Fagot-Bouquet, R. Audigier, Y. Dhome, and F. Lerasle, "Improving multi-frame data association with sparse representations for robust near-online multi-object tracking," in *European Conference on Computer Vision (ECCV)*, 2016, pp. 774–790, Springer, Cham, DOI:10.1007/978-3-319-46484-8_47.

[47] R.-E. Fan, K.-W. Chang, C.-J. Hsieh, X.-R. Wang, and C.-J. Lin, "Liblinear: A library for large linear classification," *Journal of machine learning research*, vol. 9, no. Aug, pp. 1871–1874, 2008.

[48] M. Felsberg, "Enhanced distribution field tracking using channel representations," in *IEEE International Conference on Computer Vision Workshop (ICCVW)*, 2013, pp. 121–128, IEEE, Sydney, DOI:10.1109/ICCVW.2013.22.

[49] P. Felzenszwalb, R. Girshick, D. McAllester, and D. Ramanan, "Object Detection with Discriminatively Trained Part-Based Models," *IEEE Transactions on Pattern Analysis and Machine Intelligence (PAMI)*, vol. 32, pp. 1627–1645, Sept 2010, IEEE, DOI:10.1109/TPAMI.2009.167.

[50] P. Felzenszwalb, D. McAllester, and D. Ramanan, "A discriminatively trained, multiscale, deformable part model," in *IEEE Conference on Computer Vision and Pattern Recognition (CVPR)*, June 2008, pp. 1–8, IEEE, Anchorage, DOI:10.1109/CVPR.2008.4587597.

[51] V. Franc, "Library for quadratic programming." http://cmp.felk.cvut.cz/~xfrancv/libqp/html/.

[52] L. Fridman, "Lecture notes in deep learning for self-driving cars," Jan. 2018. https://selfdrivingcars.mit.edu.

[53] C. Fries and H. J. Wuensche, "Autonomous convoy driving by night: The vehicle tracking system," in *IEEE International Conference on Technologies for Practical Robot Applications (TePRA)*, May 2015, pp. 1–6, IEEE, Woburn, DOI: 10.1109/TePRA.2015.7219675.

[54] A. Gaidon and E. Vig, "Online domain adaptation for multi-object tracking," in *CoRR*, vol. abs/1508.00776, 2015, arXiv. http://arxiv.org/abs/1508.00776.

[55] J. J. Gibson, "The perception of the visual world," 1950, Houghton Mifflin, Boston.

[56] J. Golson, "Tesla's autopilot system is reportedly getting more sensors," Aug. 2016. http://www.theverge.com/2016/8/11/12443310/tesla-autopilot-next-generation-radar-triple-camera/.

[57] J. Gorski, F. Pfeuffer, and K. Klamroth, "Biconvex sets and optimization with biconvex functions: a survey and extensions," *Mathematical methods of operations research*, vol. 66, no. 3, pp. 373–407, 2007, Springer, Berlin.

[58] H. Grabner, M. Grabner, and H. Bischof, "Real-time tracking via on-line boosting," in *British Machine Vision Conference (BMVC)*, 2006, pp. 47–56, BMVA Press, Durham, DOI:10.5244/C.20.6.

[59] H. Grabner, J. Matas, L. V. Gool, and P. Cattin, "Tracking the invisible: Learning where the object might be," in *IEEE Conference on Computer Vision and Pattern Recognition (CVPR)*, June 2010, pp. 1285–1292, IEEE, San Francisco, DOI: 10.1109/CVPR.2010.5539819.

[60] H. Grabner, C. Leistner, and H. Bischof, "Semi-supervised on-line boosting for robust tracking," in *European Conference on Computer Vision (ECCV)*, 2008, pp. 234–247, Springer, Berlin, DOI:10.1007/978-3-540-88682-2_19.

[61] J. Gräter, W. Tian, M. Lauer, and C. Stiller, "ABALID: Abbiegeassistenz mit 3D-LIDAR-Sensorik. Abschlussbericht des Teilvorhabens 3D-Objekterkennung und semantische Analyse," Karlsruhe Institute of Technology, Tech. Rep., 2016, DOI:10.2314/GBV:87492393X.

[62] E. Guizzo, "How google's self-driving car works," *IEEE Spectrum Online*, Oct. 2011. http://spectrum.ieee.org/automaton/robotics/artificial-intelligence/how-google-self-driving-car-works.

[63] H. Harb, A. Makhoul, S. Tawbi, and R. Couturier, "Comparison of different data aggregation techniques in distributed sensor networks," *Heterogeneous Crowdsourced Data Analytics*, vol. 5, pp. 4250–4263, Mar. 2017, IEEE, DOI:10.1109/ACCESS.2017.2681207.

[64] L. Hardesty, "Depth-sensing imaging system can peer through fog," 2018, MIT News, Massachusetts.

[65] S. Hare, S. Golodetz, A. Saffari, V. Vineet *et al.*, "Struck: Structured output tracking with kernels," *IEEE Transactions on Pattern Analysis and Machine Intelligence (PAMI)*, vol. 38, no. 10, pp. 2096–2109, Oct 2016, IEEE, DOI:10.1109/TPAMI.2015.2509974.

[66] S. Hare, A. Saffari, and P. H. S. Torr, "Struck: Structured output tracking with kernels," in *IEEE International Conference on Computer Vision (ICCV)*, Nov 2011, pp. 263–270, IEEE, Barcelona, DOI:10.1109/ICCV.2011.6126251.

[67] J. Hariyono, V.-D. Hoang, and K.-H. Jo, "Motion segmentation using optical flow for pedestrian detection from moving vehicle," in *International Conference on Computational Collective Intelligence*, 2014, pp. 204–213, Springer, Berlin.

[68] C. Harris and M. Stephens, "A combined corner and edge detector," in *Alvey Vision Conference*, 1988, pp. 23.1–23.6, Alvety Vision Club, DOI:10.5244/C.2.23.

[69] S. Hasinoff, D. Sharlet, R. Geiss, A. Adams *et al.*, "Burst photography for high dynamic range and low-light imaging on mobile cameras," *ACM Transactions on Graphics*, pp. 192:1–192:12, 2016, ACM, New York, DOI:10.1145/2980179.2980254.

[70] J. F. Henriques, R. Caseiro, P. Martins, and J. Batista, "Exploiting the circulant structure of tracking-by-detection with kernels," *European Conference on Computer Vision (ECCV)*, pp. 702–715, 2012, Springer, Berlin, DOI:10.1007/978-3-642-33765-9_50.

[71] J. F. Henriques, R. Caseiro, P. Martins, and J. Batista, "High-speed tracking with kernelized correlation filters," *IEEE Transactions on Pattern Analysis and Machine Intelligence (PAMI)*, vol. 37, no. 3, pp. 583–596, 2015, IEEE, DOI: 10.1109/TPAMI.2014.2345390.

[72] R. Henschel, L. Leal-Taixé, D. Cremers, and B. Rosenhahn, "Improvements to frank-wolfe optimization for multi-detector multi-object tracking," in *CoRR*, vol. abs/1705.08314, 2017, arXiv. http://arxiv.org/abs/1705.08314.

[73] Z. Hong, Z. Chen, C. Wang, X. Mei, D. Prokhorov, and D. Tao, "Multi-store tracker (muster): A cognitive psychology inspired approach to object tracking," in *IEEE Conference on Computer Vision and Pattern Recognition (CVPR)*, June 2015, pp. 749–758, IEEE, Boston, DOI:10.1109/CVPR.2015.7298675.

[74] A. W. Ingleton, "The rank of circulant matrices," *Journal of the London Mathematical Society*, vol. 1, no. 4, pp. 445–460, 1956, Wiley Online Library, Hoboken.

[75] J. Jin, A. Dundar, J. Bates, C. Farabet, and E. Culurciello, "Tracking with deep neural networks," in *Conference on Information Sciences and Systems (CISS)*, 2013, pp. 1–5, IEEE, Baltimore, DOI:10.1109/CISS.2013.6552287.

[76] B. Kapitaniak, M. Walczak, M. Kosobudzki, Z. Józwiak, and A. Bortkiewicz, "Application of eye-tracking in drivers testing: A review of research," *International Journal of Occupational Medicine and Environmental Health*, pp. 941–954, 2015, HighBeam Research, Bethesda, DOI:10.13075/ijomeh.1896.00317.

[77] M. Kaschke, K. H. Donnerhacke, and M. S. Rill, "Optical devices in ophthalmology and optometry: Technology, design principles, and clinical applications," 2014, Wiley-VCH Verlag GmbH & Co. KGaA, Weinheim.

[78] M. Kass, A. Witkin, and D. Terzopoulos, "Snakes: Active contour models," *International Journal of Computer Vision (IJCV)*, vol. 1, no. 4, pp. 321–331, 1988, Springer, Berlin, DOI:10.1007/BF00133570.

[79] R. Kasturi, D. Goldgof, P. Soundararajan, V. Manohar *et al.*, "Framework for performance evaluation of face, text, and vehicle detection and tracking in video: Data, metrics, and protocol," *IEEE Transactions on Pattern Analysis and Machine Intelligence (PAMI)*, vol. 31, no. 2, pp. 319–336, 2009, IEEE, DOI:10.1109/TPAMI.2008.57.

[80] A. Kesting, M. Treiber, M. Schönhof, and D. Helbing, "Adaptive cruise control design for active congestion avoidance," *Transportation Research Part C: Emerging Technologies*, vol. 16, no. 6, pp. 668 – 683, 2008, Elsevier, Amsterdam.

[81] M. Keuper, S. Tang, Z. Yu, B. Andres, T. Brox, and B. Schiele, "A multi-cut formulation for joint segmentation and tracking of multiple objects," in *CoRR*, vol. abs/1607.06317, 2016, arXiv. http://arxiv.org/abs/1607.06317.

[82] H. Kiani Galoogahi, T. Sim, and S. Lucey, "Multi-channel correlation filters," in *IEEE International Conference on Computer Vision (ICCV)*, 2013, pp. 3072–3079, IEEE, Sydney, DOI: 10.1109/ICCV.2013.381.

[83] H. Kiani Galoogahi, T. Sim, and S. Lucey, "Correlation filters with limited boundaries," in *IEEE Conference on Computer Vision and Pattern Recognition (CVPR)*, 2015, pp. 4630–4638, IEEE, Boston, DOI:10.1109/CVPR.2015.7299094.

[84] H. Kieritz, S. Becker, W. Hübner, and M. Arens, "Online multi-person tracking using integral channel features," in *IEEE International Conference on Advanced Video and Signal Based Surveillance (AVSS)*, Aug 2016, pp. 122–130, IEEE, Colorado Springs, DOI:10.1109/AVSS.2016.7738059.

[85] C. Kim, F. Li, A. Ciptadi, and J. M. Rehg, "Multiple hypothesis tracking revisited," in *IEEE International Conference on Computer Vision (ICCV)*, 2015, pp. 4696–4704, IEEE, Santiago, DOI: 10.1109/ICCV.2015.533.

[86] S.-Y. Kim, S.-Y. Oh, J.-K. Kang, Y. Ryu *et al.*, "Front and rear vehicle detection and tracking in the day and night times using vision and sonar sensor fusion," in *IEEE International Conference on Intelligent Robots and Systems (IROS)*, 2005, IEEE, Edmonton, DOI: 10.1109/IROS.2005.1545321.

[87] Y. Koshiba and S. Abe, "Comparison of L1 and L2 support vector machines," in *International Joint Conference on Neural Networks*, 2003, pp. 2054–2059, IEEE, Portland, DOI:10.1109/IJCNN.2003.1223724.

[88] M. Kristan, A. Leonardis, J. Matas, and et al., "The visual object tracking vot-tir2016 challenge results," in *European Conference on Computer Vision (ECCV)*, 2016, pp. 824–849, Springer, Cham, DOI:10.1007/978-3-319-48881-3_55.

[89] M. Kristan, J. Perš, V. Sulič, and S. Kovačič, "A graphical model for rapid obstacle image-map estimation from unmanned surface vehicles," in *Asian Conference on Computer Vision (ACCV)*, 2014, pp. 391–406, Springer, Cham, DOI:10.1007/978-3-319-16808-1_27.

[90] M. Kristan, R. Pflugfelder, A. Leonardis, J. Matas *et al.*, "The visual object tracking vot2016 challenge results," in *European Conference on Computer Vision Workshops (ECCVW)*, 2016, pp. 777–823, Springer, Cham, DOI:10.1007/978-3-319-48881-3_54.

[91] H. W. Kuhn, "The hungarian method for the assignment problem," *Naval Research Logistics (NRL)*, vol. 2, no. 1-2, pp. 83–97, 1955, Wiley Online Library, Hoboken.

[92] J. Y. Kwak, B. C. Ko, and J. Y. Nam, "Pedestrian tracking using online boosted random ferns learning in far-infrared imagery for safe driving at night," *IEEE Transactions on Intelligent Transportation Systems*, vol. 18, no. 1, pp. 69–81, Jan 2017, IEEE, DOI:10.1109/TITS.2016.2569159.

[93] J. Kwon and K. M. Lee, "Tracking of a non-rigid object via patch-based dynamic appearance modeling and adaptive basin hopping monte carlo sampling," in *IEEE Conference on Computer Vision and Pattern Recognition (CVPR)*, 2009, pp. 1208–1215, IEEE, Miami, DOI:10.1109/CVPR.2009.5206502.

[94] B. Lee, E. Erdenee, S. Jin, M. Y. Nam, Y. G. Jung, and P. K. Rhee, "Multi-class multi-object tracking using changing point detection," in *European Conference on Computer Vision (ECCV)*, 2016, pp. 68–83, Springer, Cham, DOI:10.1007/978-3-319-48881-3_6.

[95] P. Lenz, A. Geiger, and R. Urtasun, "Followme: Efficient online min-cost flow tracking with bounded memory and computation," in *IEEE International Conference on Computer Vision (ICCV)*, 2015, pp. 4364–4372, IEEE, Santiago, DOI:10.1109/ICCV.2015.496.

[96] E. Levinkov, S. Tang, E. Insafutdinov, and B. Andres, "Joint graph decomposition and node labeling by local search," in *CoRR*, vol. abs/1611.04399, 2016, arXiv. http://arxiv.org/abs/1611.04399.

[97] H. Li, Y. Li, and F. Porikli, "Deeptrack: Learning discriminative feature representations online for robust visual tracking," in *IEEE Transactions on Image Processing*, vol. 25, no. 4, April 2016, pp. 1834–1848, IEEE, DOI:10.1109/TIP.2015.2510583.

[98] Y. Li, Y. Zhang, Y. Xu, J. Wang, and Z. Miao, "Robust scale adaptive kernel correlation filter tracker with hierarchical convolutional features," *IEEE Signal Processing Letters*, vol. 23, no. 8, pp. 1136–1140, 2016, IEEE, DOI:10.1109/LSP.2016.2582783.

[99] Y. Li, J. Zhu, and S. C. Hoi, "Reliable patch trackers: Robust visual tracking by exploiting reliable patches," in *IEEE Conference on Computer Vision and Pattern Recognition (CVPR)*, 2015, pp. 353–361, IEEE, Boston, DOI:10.1109/CVPR.2015.7298632.

[100] T. Liu, G. Wang, and Q. Yang, "Real-time part-based visual tracking via adaptive correlation filters," in *IEEE Conference on Computer Vision and Pattern Recognition (CVPR)*, 2015, pp. 4902–4912, IEEE, Boston, DOI:10.1109/CVPR.2015.7299124.

[101] D. G. Lowe, "Distinctive image features from scale-invariant keypoints," in *IEEE International Conference on Computer Vision (ICCV)*, vol. 60, no. 2, 2004, pp. 91–110, Kluwer Academic Publishers, Dordrecht, DOI:10.1023/B:VISI.0000029664.99615.94.

[102] B. D. Lucas and T. Kanade, "An iterative image registration technique with an application to stereo vision," *International Joint Conference on Artificial Intelligence (IJCAI)*, vol. 2, pp. 647–679, 1981, Morgan Kaufmann Publishers, San Francisco.

[103] A. Lukežič, T. Vojíř, L. Čehovin, J. Matas, and M. Kristan, "Discriminative correlation filter with channel and spatial reliability," in *IEEE Conference on Computer Vision and Pattern Recognition (CVPR)*, 2016, IEEE, Honolulu, DOI:10.1109/CVPR.2017.515.

[104] S. Lyu, M.-C. Chang, D. Du, L. Wen *et al.*, "Ua-detrac 2017: Report of avss2017 & it4s challenge on advance traffic monitoring," in *IEEE International Conference on Advanced Video and Signal based Surveillance (AVSS)*, Aug 2017, IEEE, Lecce, DOI:10.1109/AVSS.2017.8078560.

[105] C. Ma, J.-B. Huang, X. Yang, and M.-H. Yang, "Hierarchical convolutional features for visual tracking," in *IEEE International Conference on Computer Vision (ICCV)*, 2015, IEEE, Santiago, DOI:10.1109/ICCV.2015.352.

[106] E. Maggio and A. Cavallaro, "Multi-part target representation for color tracking," in *IEEE International Conference on Image Processing*, vol. 1, Sept 2005, pp. I–729–32, IEEE, Genova, DOI:10.1109/ICIP.2005.1529854.

[107] H. Malm, M. Oskarsson, E. Warrant, P. Clarberg, J. Hasselgren, and C. Lejdfors, "Adaptive enhancement and noise reduction in very low light-level video," in *IEEE International Conference on Computer Vision (ICCV)*, Oct 2007, pp. 1–8, IEEE, Rio de Janeiro, DOI:10.1109/ICCV.2007.4409007.

[108] X. Mei and H. Ling, "Robust visual tracking using L1 minimization," in *IEEE Conference on Computer Vision and Pattern Recognition (CVPR)*, 2009, pp. 1436–1443, IEEE, Kyoto, DOI:10.1109/ICCV.2009.5459292.

[109] X. Mei, H. Ling, Y. Wu, E. Blasch, and L. Bai, "Minimum error bounded efficient L1 tracker with occlusion detection," in *IEEE Conference on Computer Vision and Pattern Recognition (CVPR)*, 2011, pp. 1257–1264, IEEE, Colorado Springs, DOI:10.1109/CVPR.2011.5995421.

[110] A. Milan, L. Leal-Taixé, I. Reid, S. Roth, and K. Schindler, "MOT16: A benchmark for multi-object tracking," in *CoRR*, vol. abs/1603.00831, Mar. 2016, arXiv. http://arxiv.org/abs/1603.00831.

[111] A. Milan, S. Roth, and K. Schindler, "Continuous energy minimization for multitarget tracking," *IEEE Transactions on Pattern Analysis and Machine Intelligence (PAMI)*, vol. 36, no. 1, pp. 58–72, 2014, IEEE, DOI:10.1109/TPAMI.2013.103.

[112] A. Milan, K. Schindler, and S. Roth, "Detection-and trajectory-level exclusion in multiple object tracking," in *IEEE Conference on Computer Vision and Pattern Recognition (CVPR)*, 2013, pp. 3682–3689, IEEE, Portland, DOI:10.1109/CVPR.2013.472.

[113] M. Mueller, N. Smith, and B. Ghanem, "Context-aware correlation filter tracking," in *IEEE Conference on Computer Vision and Pattern Recognition (CVPR)*, 2017, pp. 1396–1404, IEEE, Honolulu, DOI:10.1109/CVPR.2017.152.

[114] H. Nam and B. Han, "Learning multi-domain convolutional neural networks for visual tracking," *IEEE Conference on Computer Vision and Pattern Recognition (CVPR)*, pp. 4293–4302, 2015, IEEE, Las Vegas, DOI:10.1109/CVPR.2016.465.

[115] G. Nebehay and R. Pflugfelder, "Clustering of static-adaptive correspondences for deformable object tracking," in *IEEE Conference on Computer Vision and Pattern Recognition (CVPR)*, 2015, pp. 2784–2791, IEEE, Boston, DOI:10.1109/CVPR.2015.7298895.

[116] P. Niyogi, F. Girosi, and T. Poggio, "Incorporating prior information in machine learning by creating virtual examples," *Proceedings of the IEEE*, vol. 86, no. 11, pp. 2196–2209, 1998, IEEE, DOI:10.1109/5.726787.

[117] T. Ojala, M. Pietikainen, and T. Maenpaa, "Multiresolution gray-scale and rotation invariant texture classification with local binary patterns," *IEEE Transactions on Pattern Analysis and Machine Intelligence (PAMI)*, pp. 971–987, Jul 2002, IEEE, DOI:10.1109/TPAMI.2002.1017623.

[118] R. O'Malley, E. Jones, and M. Glavin, "Rear-lamp vehicle detection and tracking in low-exposure color video for night conditions," *IEEE Transactions on Intelligent Transportation Systems*, vol. 11, no. 2, pp. 453–462, June 2010, IEEE, DOI:10.1109/TITS.2010.2045375.

[119] S. Oron, A. Bar-Hillel, D. Levi, and S. Avidan, "Locally orderless tracking," in *IEEE Conference on Computer Vision and Pattern Recognition (CVPR)*, 2012, pp. 1940–1947, IEEE, DOI:10.1109/CVPR.2012.6247895.

[120] A. Ošep, W. Mehner, M. Mathias, and B. Leibe, "Combined image-and world-space tracking in traffic scenes," in *IEEE International Conference on Robotics and Automation (ICRA)*, 2017, pp. 1988–1995, IEEE, DOI:10.1109/ICRA.2017.7989230.

[121] E. Park, W. Liu, O. Russakovsky, J. Deng, F.-F. Li, and A. Berg, "Large scale visual recognition challenge (ilsvrc) 2017 overview," in *IEEE International Conference on Computer Vision Workshop (ICCVW)*, 2017. http://image-net.org/challenges/LSVRC/2017/.

[122] S. Pellegrini, A. Ess, K. Schindler, and L. van Gool, "You'll never walk alone: Modeling social behavior for multi-target tracking," in *IEEE International Conference on Computer Vision (ICCV)*, Sept 2009, pp. 261–268, IEEE, Kyoto, DOI:10.1109/ICCV.2009.5459260.

[123] S. Petit, "World vehicle population rose 4.6% in 2016," *WardsAuto*, Oct. 2017. http://wardsauto.com/analysis/world-vehicle-population-rose-46-2016.

[124] H. Pirsiavash, D. Ramanan, and C. C. Fowlkes, "Globally-optimal greedy algorithms for tracking a variable number of objects," in *IEEE Conference on Computer Vision and Pattern Recognition (CVPR)*, 2011, pp. 1201–1208, IEEE, Colorado Springs, DOI:10.1109/CVPR.2011.5995604.

[125] C. Rao, C. Yao, X. Bai, W. Qiu, and W. Liu, "Online random ferns for robust visual tracking," in *International Conference on Pattern Recognition (ICPR)*, 2012, pp. 1447–1450, IEEE, Tsukuba, ISSN:1051-4651.

[126] E. Ristani and C. Tomasi, "Tracking multiple people online and in real time," in *Asian Conference on Computer Vision (ACCV)*, 2014, pp. 444–459, Springer, Cham, DOI:10.1007/978-3-319-16814-2_29.

[127] K. Robert, "Night-time traffic surveillance: A robust framework for multi-vehicle detection, classification and tracking," in *IEEE International Conference on Advanced Video and Signal Based Surveillance (AVSS)*, Sept 2009, pp. 1–6, IEEE, Genova, DOI:10.1109/AVSS.2009.98.

[128] A. Roshan Zamir, A. Dehghan, and M. Shah, "Gmcp-tracker: Global multi-object tracking using generalized minimum clique graphs," in *European Conference on Computer Vision (ECCV)*, 2012, pp. 343–356, Springer, Berlin, DOI:10.1007/978-3-642-33709-3_25.

[129] D. A. Ross, J. Lim, R.-S. Lin, and M.-H. Yang, "Incremental learning for robust visual tracking," *International Journal of Computer Vision (IJCV)*, vol. 77, no. 1-3, pp. 125–141, 2008, Springer, Berlin, DOI:10.1007/s11263-007-0075-7.

[130] E. Rosten and T. Drummond, "Machine learning for high-speed corner detection," in *European Conference on Computer Vision (ECCV)*, 2006, pp. 430–443, Springer, Berlin, DOI:10.1007/11744023_34.

[131] A. Sadeghian, A. Alahi, and S. Savarese, "Tracking the untrackable: Learning to track multiple cues with long-term dependencies," in *IEEE Conference on Computer Vision and Pattern Recognition (CVPR)*, 2017, pp. 300–311, IEEE, Venice, DOI:10.1109/ICCV.2017.41.

[132] A. Saffari, C. Leistner, J. Santner, M. Godec, and H. Bischof, "On-line random forests," in *IEEE International Conference on Computer Vision Workshops (ICCV)*, Sept 2009, pp. 1393–1400, IEEE, Kyoto, DOI:10.1109/ICCVW.2009.5457447.

[133] R. Sanchez-Matilla, F. Poiesi, and A. Cavallaro, "Online multi-target tracking with strong and weak detections," in *European Conference on Computer Vision Workshop (ECCVW)*, 2016, pp. 84–99, Springer, Cham, DOI:10.1007/978-3-319-48881-3_7.

[134] B. Schölkopf, R. Herbrich, and A. J. Smola, "A generalized representer theorem," in *International conference on computational learning theory*, 2001, pp. 416–426, Springer, Berlin, DOI:10.1007/3-540-44581-1_27.

[135] B. Schölkopf, A. J. Smola, F. Bach *et al.*, *Learning with kernels: support vector machines, regularization, optimization, and beyond.* MIT press, 2002.

[136] S. Shalev-Shwartz, Y. Singer, N. Srebro, and A. Cotter, "Pegasos: primal estimated sub-gradient solver for svm," *Mathematical Programming*, vol. 127, no. 1, pp. 3–30, Mar 2011, springer, DOI:10.1007/s10107-010-0420-4.

[137] B. Shen and J.-S. Fu, "A method of data aggregation for wearable sensor systems," *Sensors*, 2016, NCBI, Bethesda, DOI:10.3390/s16070954.

[138] A. W. M. Smeulders, D. M. Chu, R. Cucchiara, S. Calderara, A. Dehghan, and M. Shah, "Visual tracking: An experimental survey," *IEEE Transactions on Pattern Analysis and Machine Intelligence (PAMI)*, vol. 36, no. 7, pp. 1442–1468, July 2014, IEEE, DOI:10.1109/TPAMI.2013.230.

[139] S. Stalder, H. Grabner, and L. v. Gool, "Beyond semi-supervised tracking: Tracking should be as simple as detection, but not simpler than recognition," in *IEEE International Conference on Computer Vision Workshop (ICCVW)*, Sept 2009, pp. 1409–1416, IEEE, Kyoto, DOI:10.1109/ICCVW.2009.5457445.

[140] J. Stanley and B. Steinhardt, "Bigger monster, weaker chains: The growth of an american surveillance society," *Ethics and Emerging Technologies*, pp. 269–284, 2014, Palgrave Macmillan UK.

[141] J. S. Supancic and D. Ramanan, "Self-paced learning for long-term tracking," in *IEEE Conference on Computer Vision and Pattern Recognition (CVPR)*, 2013, pp. 2379–2386, IEEE, Portland, DOI:10.1109/CVPR.2013.308.

[142] S. Tang, B. Andres, M. Andriluka, and B. Schiele, "Subgraph decomposition for multi-target tracking," in *IEEE Conference on Computer Vision and Pattern Recognition (CVPR)*, June 2015, pp. 5033–5041, IEEE, Boston, DOI:10.1109/CVPR.2015.7299138.

[143] Y. Tang, "Deep learning using linear support vector machines," in *CoRR*, vol. abs/1306.0239, 2013, arXiv. http://arxiv.org/abs/1306.0239.

[144] W. Tian, L. Chen, K. Zou, and M. Lauer, "Vehicle tracking at nighttime by kernelized experts with channel-wise and temporal reliability estimation," *IEEE Transactions on Intelligent Transportation Systems*, vol. 19, no. 10, pp. 3159–3169, Dec 2017, IEEE, DOI:10.1109/TITS.2017.2771410.

[145] W. Tian and M. Lauer, "Joint tracking with event grouping and temporal constraints," in *IEEE International Conference on Advanced Video and Signal Based Surveillance (AVSS)*, Aug. 2017, pp. 1–5, IEEE, Lecce, DOI:10.1109/AVSS.2017.8078515.

[146] W. Tian and M. Lauer, "Tracking vulnerable road users with severe occlusion by adaptive part filter modeling," in *IEEE International Conference on Vehicular Electronics and Safety (ICVES)*, June 2017, pp. 139–144, IEEE, Vienna, DOI:10.1109/ICVES.2017.7991915.

[147] W. Tian, M. Lauer, and L. Chen, "Online multi-object tracking using joint domain information in traffic scenarios," *IEEE Transactions on Intelligent Transportation Systems*, pp. 1–11, Jan 2019, IEEE, DOI:10.1109/TITS.2019.2892413.

[148] W. Tian and M. Lauer, "Fast and Robust Cyclist Detection for Monocular Camera Systems," in *Doctor Consortium at International joint Conference on Computer Vision, Imaging and Computer Graphics Theory and Applications (VISIGRAPP)*, March 2015, pp. 3–8, SCITEPRESS, Berlin.

[149] W. Tian and M. Lauer, "Fast Cyclist Detection by Cascaded Detector and Geometric Constraint," in *IEEE Conference on Intelligent Transportation Systems (ITSC)*, Sept 2015, pp. 1286–1291, IEEE, Las Palmas, DOI:10.1109/ITSC.2015.211.

[150] W. Tian and M. Lauer, "Detection and orientation estimation for cyclists by max pooled features," in *International Joint Conference on Computer Vision, Imaging and Computer Graphics Theory and Applications (VISIGRAPP)*, 2017, pp. 17–26, SCITEPRESS, Porto, DOI: 10.5220/0006085500170026.

[151] W. Tian and M. Lauer, "Tracking objects with severe occlusion by adaptive part filter modeling - in traffic scenes and beyond," *IEEE Intelligent Transportation Systems Magazine*, pp. 60–73, Oct. 2018, IEEE, DOI:10.1109/MITS.2018.2867517.

[152] E. Tingwall, "Sensory overload: How the new mercedes s-class sees all," *Car and Driver*, 2013. http://blog.caranddriver.com/sensory-overload-how-the-new-mercedes-s-class-sees-all/.

[153] C. Tomasi and T. Kanade, "Detection and tracking of point features," 1991, School of Computer Science, Carnegie Mellon Univ. Pittsburgh.

[154] M. Villamizar, J. Andrade-Cetto, A. Sanfeliu, and F. Moreno-Noguer, "Boosted random ferns for object detection," *IEEE Transaction on Pattern Analysis and Machine Intelligence (PAMI)*, vol. 40, no. 2, pp. 272–288, 2018, IEEE, DOI:10.1109/TPAMI.2017.2676778.

[155] P. Viola and M. Jones, "Robust Real-Time Face Detection," *International Journal of Computer Vision (IJCV)*, no. 2, pp. 137–154, 2004, Springer, Berlin, DOI:10.1023/B:VISI.0000013087.49260.fb.

[156] S. Wang and C. C. Fowlkes, "Learning optimal parameters for multi-target tracking with contextual interactions," *International Journal of Computer Vision (IJCV)*, pp. 1–18, 2016, Springer, Berlin, DOI:10.1007/s11263-016-0960-z.

[157] N. Wax, "Signal-to-noise improvement and the statistics of track populations," *Journal of Applied Physics*, vol. 26, no. 5, pp. 586–595, 1955, American Institute of Physics, Maryland, DOI:10.1063/1.1722046.

[158] P. Weinzaepfel, J. Revaud, Z. Harchaoui, and C. Schmid, "Deepflow: Large displacement optical flow with deep matching," in *IEEE International Conference on Computer Vision (ICCV)*, 2013, pp. 1385–1392, IEEE, Sydney, DOI:10.1063/1.1722046.

[159] L. Wen, D. Du, Z. Cai, Z. Lei *et al.*, "UA-DETRAC: A new benchmark and protocol for multi-object tracking," in *CoRR*, vol. abs/1511.04136, 2015, arXiv. http://arxiv.org/abs/1511.04136.

[160] L. Wen, Z. Lei, S. Lyu, S. Z. Li, and M. Yang, "Exploiting hierarchical dense structures on hypergraphs for multi-object tracking," *IEEE Transaction on Pattern Analysis and Machine Intelligence (PAMI)*, vol. 38, no. 10, pp. 1983–1996, 2016, IEEE, DOI:10.1109/TPAMI.2015.2509979.

[161] L. Wen, P. Zhu, D. Du, X. Bian *et al.*, "Visdrone-sot2018: The vision meets drone single-object tracking challenge results," in *European Conference on Computer Vision Workshop (ECCVW)*, 2018, pp. 469–495, Springer, Cham, DOI:10.1007/978-3-030-11021-5_28.

[162] S. A. Wibowo, H. Lee, E. K. Kim, and S. Kim, "Fast generative approach based on sparse representation for visual tracking," in *Joint International Conference on Soft Computing and Intelligent Systems (SCIS) and International Symposium on Advanced Intelligent Systems*, 2016, pp. 778–783, IEEE, Sapporo, DOI:10.1109/SCIS-ISIS.2016.0169.

[163] World Health Organization, "Global status report on road safety 2015," 2015, WHO, Geneva.

[164] C. R. Wren, A. Azarbayejani, T. Darrell, and A. P. Pentland, "Pfinder: Real-time tracking of the human body," *IEEE Transaction on Pattern Analysis and Machine Intelligence (PAMI)*, vol. 19, no. 7, pp. 780–785, 1997, IEEE, DOI:10.1109/34.598236.

[165] J. Wright, A. Y. Yang, A. Ganesh, S. S. Sastry, and Y. Ma, "Robust face recognition via sparse representation," *IEEE Transaction on Pattern Analysis and Machine Intelligence (PAMI)*, vol. 31, no. 2, pp. 210–227, 2009, IEEE, DOI: 10.1109/TPAMI.2008.79.

[166] Y. Wu, J. Lim, and M. H. Yang, "Object tracking benchmark," *IEEE Transactions on Pattern Analysis and Machine Intelligence (PAMI)*, vol. 37, no. 9, pp. 1834–1848, Sept 2015, IEEE, DOI:10.1109/TPAMI.2014.2388226.

[167] Y. Wu, J. Lim, and M.-H. Yang, "Online object tracking: A benchmark," in *IEEE Conference on Computer Vision and Pattern Recognition (CVPR)*, 2013, pp. 2411–2418, IEEE, Portland, DOI:10.1109/CVPR.2013.312.

[168] Y. Xiang, A. Alahi, and S. Savarese, "Learning to track: Online multi-object tracking by decision making," in *IEEE International Conference on Computer Vision (ICCV)*, 2015, pp. 4705–4713, IEEE, Santiago, DOI:10.1109/ICCV.2015.534.

[169] J. H. Yoon, C. R. Lee, M. H. Yang, and K. J. Yoon, "Online multi-object tracking via structural constraint event aggregation," in *IEEE Conference on Computer Vision and Pattern Recognition (CVPR)*, June 2016, pp. 1392–1400, IEEE, Las Vegas, DOI:10.1109/CVPR.2016.155.

[170] J. H. Yoon, M.-H. Yang, J. Lim, and K.-J. Yoon, "Bayesian multi-object tracking using motion context from multiple objects," in *IEEE Winter Conference on Applications of Computer Vision (WACV)*, 2015, pp. 33–40, IEEE, Waikoloa, DOI:10.1109/WACV.2015.12.

[171] H. Zhang, A. Geiger, and R. Urtasun, "Understanding high-level semantics by modeling traffic patterns," in *IEEE International Conference on Computer Vision and Pattern Recognition (CVPR)*, 2013, pp. 3056–3063, IEEE, Sydney, DOI:10.1109/ICCV.2013.379.

[172] J. Zhang, S. Ma, and S. Sclaroff, "MEEM: robust tracking via multiple experts using entropy minimization," in *European Conference on Computer Vision (ECCV)*, 2014, pp. 188–203, Springer, Cham, DOI:10.1007/978-3-319-10599-4_13.

[173] T. Zhang, B. Ghanem, S. Liu, and N. Ahuja, "Low-rank sparse learning for robust visual tracking," in *European Conference on Computer Vision (ECCV)*, 2012, pp. 470–484, Springer, Berlin, DOI:10.1007/978-3-642-33783-3_34.

[174] P. Zhu, L. Wen, X. Bian, L. Haibin, and Q. Hu, "Vision meets drones: A challenge," *CoRR*, vol. abs/1804.07437, 2018, arXiv. http://arxiv.org/abs/1804.07437.

[175] P. Zhu, L. Wen, D. Du, X. Bian *et al.*, "Visdrone-vdt2018: The vision meets drone video detection and tracking challenge results," in *European Conference on Computer Vision Workshop (ECCVW)*, 2018, pp. 496–518, Springer, Cham, DOI:10.1007/978-3-030-11021-5_29.

[176] Q. Zou, H. Ling, S. Luo, Y. Huang, and M. Tian, "Robust nighttime vehicle detection by tracking and grouping headlights," *IEEE Transactions on Intelligent Transportation Systems*, vol. 16, no. 5, pp. 2838–2849, Oct 2015, IEEE, DOI:10.1109/TITS.2015.2425229.

[177] W. Zuo, X. Wu, L. Lin, L. Zhang, and M. Yang, "Learning support correlation filters for visual tracking," in *IEEE Transaction on Pattern Analysis and Machine Intelligence (PAMI)*, 2016, pp. 1–14, IEEE, DOI:10.1109/TPAMI.2018.2829180.

Schriftenreihe
Institut für Mess- und Regelungstechnik
Karlsruher Institut für Technologie
(1613-4214)

Band 001 Hans, Annegret
Entwicklung eines Inline-Viskosimeters auf Basis eines magnetisch-induktiven Durchflussmessers. 2004
ISBN 3-937300-02-3

Band 002 Heizmann, Michael
Auswertung von forensischen Riefenspuren mittels automatischer Sichtprüfung. 2004
ISBN 3-937300-05-8

Band 003 Herbst, Jürgen
Zerstörungsfreie Prüfung von Abwasserkanälen mit Klopfschall. 2004
ISBN 3-937300-23-6

Band 004 Kammel, Sören
Deflektometrische Untersuchung spiegelnd reflektierender Freiformflächen. 2005
ISBN 3-937300-28-7

Band 005 Geistler, Alexander
Bordautonome Ortung von Schienenfahrzeugen mit Wirbelstrom-Sensoren. 2007
ISBN 978-3-86644-123-1

Band 006 Horn, Jan
Zweidimensionale Geschwindigkeitsmessung texturierter Oberflächen mit flächenhaften bildgebenden Sensoren. 2007
ISBN 978-3-86644-076-0

Band 007 Hoffmann, Christian
Fahrzeugdetektion durch Fusion monoskopischer Videomerkmale. 2007
ISBN 978-3-86644-139-2

Band 008 Dang, Thao
Kontinuierliche Selbstkalibrierung von Stereokameras. 2007
ISBN 978-3-86644-164-4

Band 009 Kapp, Andreas
Ein Beitrag zur Verbesserung und Erweiterung der Lidar-Signalverarbeitung für Fahrzeuge. 2007
ISBN 978-3-86644-174-3

Band 010 Horbach, Jan
Verfahren zur optischen 3D-Vermessung spiegelnder Oberflächen. 2008
ISBN 978-3-86644-202-3

Band 011 Böhringer, Frank
Gleisselektive Ortung von Schienenfahrzeugen mit bordautonomer Sensorik. 2008
ISBN 978-3-86644-196-5

Band 012 Xin, Binjian
Auswertung und Charakterisierung dreidimensionaler Messdaten technischer Oberflächen mit Riefentexturen. 2009
ISBN 978-3-86644-326-6

Band 013 Cech, Markus
Fahrspurschätzung aus monokularen Bildfolgen für innerstädtische Fahrerassistenzanwendungen. 2009
ISBN 978-3-86644-351-8

Band 014 Speck, Christoph
Automatisierte Auswertung forensischer Spuren auf Patronenhülsen. 2009
ISBN 978-3-86644-365-5

Band 015 Bachmann, Alexander
Dichte Objektsegmentierung in Stereobildfolgen. 2010
ISBN 978-3-86644-541-3

Band 016 Duchow, Christian
Videobasierte Wahrnehmung markierter Kreuzungen mit lokalem Markierungstest und Bayes'scher Modellierung. 2011
ISBN 978-3-86644-630-4

Band 017 Pink, Oliver
Bildbasierte Selbstlokalisierung von Straßenfahrzeugen. 2011
ISBN 978-3-86644-708-0

Band 018 Hensel, Stefan
Wirbelstromsensorbasierte Lokalisierung von Schienenfahrzeugen in topologischen Karten. 2011
ISBN 978-3-86644-749-3

Band 019 Carsten Hasberg
Simultane Lokalisierung und Kartierung spurgeführter Systeme. 2012
ISBN 978-3-86644-831-5

Band 020 Pitzer, Benjamin
Automatic Reconstruction of Textured 3D Models. 2012
ISBN 978-3-86644-805-6

Band 021 Roser, Martin
Modellbasierte und positionsgenaue Erkennung von Regentropfen in Bildfolgen zur Verbesserung von videobasierten Fahrerassistenzfunktionen. 2012
ISBN 978-3-86644-926-8

Band 022 Loose, Heidi
Dreidimensionale Straßenmodelle für Fahrerassistenzsysteme auf Landstraßen. 2013
ISBN 978-3-86644-942-8

Band 023 Rapp, Holger
Reconstruction of Specular Reflective Surfaces using Auto-Calibrating Deflectometry. 2013
ISBN 978-3-86644-966-4

Band 024 Moosmann, Frank
Interlacing Self-Localization, Moving Object Tracking and Mapping for 3D Range Sensors. 2013
ISBN 978-3-86644-977-0

Band 025 Geiger, Andreas
Probabilistic Models for 3D Urban Scene Understanding from Movable Platforms. 2013
ISBN 978-3-7315-0081-0

Band 026 Hörter, Marko
Entwicklung und vergleichende Bewertung einer bildbasierten Markierungslichtsteuerung für Kraftfahrzeuge. 2013
ISBN 978-3-7315-0091-9

Band 027 Kitt, Bernd
Effiziente Schätzung dichter Bewegungsvektorfelder unter Berücksichtigung der Epipolargeometrie zwischen unterschiedlichen Ansichten einer Szene. 2013
ISBN 978-3-7315-0105-3

Band 028 Lategahn, Henning
Mapping and Localization in Urban Environments Using Cameras. 2013
ISBN 978-3-7315-0135-0

Band 029 Tischler, Karin
Informationsfusion für die kooperative Umfeldwahrnehmung vernetzter Fahrzeuge. 2014
ISBN 978-3-7315-0166-4

Band 030 Schmidt, Christian
Fahrstrategien zur Unfallvermeidung im Straßenverkehr für Einzel- und Mehrobjektszenarien. 2014
ISBN 978-3-7315-0198-5

Band 031 Firl, Jonas
Probabilistic Maneuver Recognition in Traffic Scenarios. 2014
ISBN 978-3-7315-0287-6

Band 032 Schönbein, Miriam
Omnidirectional Stereo Vision for Autonomous Vehicles. 2015
ISBN 978-3-7315-0357-6

Band 033 Nicht erschienen

Band 034 Liebner, Martin
Fahrerabsichtserkennung und Risikobewertung für warnende Fahrerassistenzsysteme. 2016
ISBN 978-3-7315-0508-2

Band 035 Ziegler, Julius
Optimale Trajektorienplanung für Automobile. 2017
ISBN 978-3-7315-0553-2

Band 036 Harms, Hannes
Genauigkeitsuntersuchung von binokularen Normalenvektoren für die Umfeldwahrnehmung. 2017
ISBN 978-3-7315-0628-7

Band 037 Ruhhammer, Christian
Inferenz von Kreuzungsinformationen aus Flottendaten. 2017
ISBN 978-3-7315-0721-5

Band 038 Stein, Denis
Mobile laser scanning based determination of railway network topology and branching direction on turnouts. 2018
ISBN 978-3-7315-0743-7

Band 039 Yi, Boliang
Integrated Planning and Control for Collision Avoidance Systems. 2018
ISBN 978-3-7315-0785-7

Band 040 Schwarze, Tobias
Compact Environment Modelling from Unconstrained Camera Platforms. 2018
ISBN 978-3-7315-0801-4

Band 041 Knorr, Moritz
Self-Calibration of Multi-Camera Systems for Vehicle Surround Sensing. 2018
ISBN 978-3-7315-0765-9

Band 042 Rabe, Johannes
Lane-Precise Localization with Production Vehicle Sensors and Application to Augmented Reality Navigation. 2018
ISBN 978-3-7315-0854-0

Band 043 Weiser, Andreas
Probabilistische Vorhersage von Fahrstreifenwechseln für hochautomatisiertes Fahren auf Autobahnen. 2019
ISBN 978-3-7315-0794-9

Band 044 Tian, Wei
Novel Aggregated Solutions for Robust Visual Tracking in Traffic Scenarios. 2019
ISBN 978-3-7315-0915-8
